Stochastic Process Optimization using Aspen Plus®

Stochastic Process Optimization using Aspen Plus®

By

Juan Gabriel Segovia-Hernández and
Fernando Israel Gómez-Castro

CRC Press
Taylor & Francis Group
Boca Raton London New York

CRC Press is an imprint of the
Taylor & Francis Group, an **informa** business

CRC Press
Taylor & Francis Group
6000 Broken Sound Parkway NW, Suite 300
Boca Raton, FL 33487-2742

CRC Press is an imprint of Taylor & Francis Group, an Informa business

Printed on acid-free paper

International Standard Book Number-13: 978-1-4987-8510-5 (Hardback)

Library of Congress Cataloging-in-Publication Data

Names: Segovia-Hernández, Juan Gabriel, author. | Gómez-Castro, Fernando Israel, author.
Title: Stochastic process optimization using Aspen plus / Juan Gabriel Segovia-Hernández and Fernando Israel Gómez-Castro.
Description: Boca Raton : Taylor & Francis, CRC Press, 2017. | Includes bibliographical references and index.
Identifiers: LCCN 2017004915 | ISBN 9781498785105 (hardback : alk. paper) | ISBN 9781498785112 (ebook)
Subjects: LCSH: Chemical processes—Data processing. | Mathematical optimization. | Stochastic processes.
Classification: LCC TP184 .S44 2017 | DDC 660/.28—dc23
LC record available at https://lccn.loc.gov/2017004915

Visit the Taylor & Francis Web site at
http://www.taylorandfrancis.com

and the CRC Press Web site at
http://www.crcpress.com

To my lovely mom.

To the memory of my dad.

Juan Gabriel Segovia-Hernández

To the reason of all my efforts, the one who keeps me strong, my dear wife Judith.

To all my family, thank you for your support: My parents, Esthela and Fernando. My sister, Ana. My brothers, Erick and Dante.

Fernando Israel Gómez-Castro

Contents

Preface ... xi
Acknowledgement ... xiii
Editors ... xv
Contributors ... xvii

1 Introduction to Optimization ... 1
1.1 What Is Optimization? ... 1
1.2 Mathematical Modeling and Optimization ... 2
1.3 Classification of Optimization Problems ... 4
1.4 Objective Function ... 5
1.5 Optimization with Constraints: Feasible Region ... 7
1.6 Multiobjective Optimization ... 8
1.6.1 Weighted Sum Method ... 11
1.6.2 Constraint Method ... 12
1.7 Process Optimization ... 12
References ... 14

2 Deterministic Optimization ... 15
2.1 Introduction ... 15
2.2 Single-Variable Deterministic Optimization ... 15
2.3 Continuity and Convexity ... 19
2.4 Unconstrained Optimization ... 24
2.5 Equality-Constrained Optimization ... 25
2.5.1 Method of Lagrange Multipliers ... 26
2.5.2 Generalized Reduced Gradient Method ... 29
2.6 Equality- and Inequality-Constrained Optimization ... 32
2.6.1 Active Set Strategy ... 33
2.7 Software for Deterministic Optimization ... 36
References ... 36

3 Stochastic Optimization ... 39
3.1 Introduction to Stochastic Optimization ... 39
3.2 Stochastic Optimization vs. Deterministic Optimization ... 40
3.3 Stochastic Optimization with Constraints ... 40
3.4 Genetic Algorithms ... 42
3.5 Differential Evolution ... 43
3.6 Tabu Search ... 45
3.7 Simulated Annealing ... 47
3.8 Other Methods ... 48
3.8.1 Ant Colony Optimization ... 49

3.8.2 Particle Swarm Optimization 50
3.8.3 Harmony Search 50
References 51

4 The Simulator Aspen Plus® 55
4.1 Importance of Software for Process Analysis 55
4.2 Characteristics of the Process Simulator Aspen Plus 56
4.3 Sequential Modular Simulation 58
References 59

5 Direct Optimization in Aspen Plus® 61
5.1 Optimization Methods 61
5.2 Sensitivity Analysis Tools in Aspen Plus 62
5.3 Sequential Quadratic Programming in Aspen Plus 62
5.4 Optimization of a Heat Exchanger 63
5.4.1 Description of the Problem 63
5.4.2 Initial Simulation 63
5.4.3 Optimization through Sensitivity Analysis 68
5.4.4 Optimization through Sequential Quadratic Programming 73
5.5 Optimization of a Flash Drum 76
5.5.1 Description of the Problem 76
5.5.2 Initial Simulation 78
5.5.3 Optimization through Sensitivity Analysis 83
5.5.4 Optimization through Sequential Quadratic Programming 85
5.6 Optimization of a Tubular Reactor 91
5.6.1 Description of the Problem 91
5.6.2 Initial Simulation 92
5.6.3 Optimization through Sensitivity Analysis 98
5.6.4 Optimization through Sequential Quadratic Programming 103
References 106

6 Optimization using Aspen Plus® and Stochastic Toolbox 107
6.1 Introduction 107
6.2 Software for Stochastic Optimization 107
6.3 Linking Aspen Plus with the Stochastic Optimization Software 109
6.3.1 Creating a Function to be Optimized with MATLAB....110
6.3.2 Creating a Subroutine in Microsoft Excel 111
6.4 Mono-Objective Optimization of a Multicomponent Distillation Column 118
6.5 Multi-Objective Optimization of a Multicomponent Distillation Column 119

6.6 Conclusions 122
References 123

7 **Using External User-Defined Block Model in Aspen Plus®** 125
7.1 Introduction 125
7.2 Importance of User-Defined Block Models 125
7.3 Previous Work and Loading a User-Defined Block Model in Aspen Plus 126
7.4 Linking User-Defined Block Model with Microsoft Excel and MATLAB 130
7.5 Conclusions 138
References 139

8 **Optimization with a User Kinetic Model** 141
8.1 Introduction 141
8.2 Kinetic Models Allowed in Aspen Plus 141
8.3 Developing a User Kinetic Model 143
8.4 Loading a User Kinetic Model in Aspen Plus 149
8.5 Optimization of a Reactive Distillation Column with a User Kinetic Model 152
8.6 Reactive Distillation Column with a Default Kinetic Model 159
8.7 Conclusions 166
References 166

9 **Optimization of a Biobutanol Production Process** 169
9.1 Introduction 169
9.2 Description of the Process 171
9.3 Thermodynamics and Kinetic Model 173
9.4 Optimization Process 186
9.5 Optimization Results 188
9.6 Conclusions 190
References 191

10 **Optimization of a Silane Production Process** 193
10.1 Introduction 193
10.2 Silane Production 195
10.3 Description of the Process Using Reactive Distillation 196
10.4 Economic Potential of Reactive Distillation Production of Silane 198
10.5 Thermodynamics and Kinetic Model 198
10.6 Initial Design 200
10.6.1 Buildup of the Initial Column 201
10.7 Process Optimization 212

10.7.1 Economic Objective Function 212
10.7.2 Environmental Objective Function 212
10.7.3 Global Stochastic Optimization 213
10.7.4 Results 213
10.8 Conclusions 216
References 217

Index 219

Preface

We optimize something every day, even without thinking about it. Each time we take a decision from among a set of possibilities, we optimize by selecting the best alternative in terms of a criterion established by ourselves. If we follow this though, we may find out that optimization problems exist even from the very beginning of mankind. But, what is optimization? A simple definition: optimization is to choose the best alternative among a set of feasible options. This can be performed in many different ways: deciding in terms of previous experiences, testing all the alternatives and choosing the best one, or using mathematical approaches. When using a mathematical optimization approach, it is important to develop reliable models which properly represent the system to be optimized. Thus, learning optimization techniques requires at least basic knowledge in modelling and mathematics.

In all engineering areas, optimization has a wide range of applications due to the large number of decision-making situations involved in an engineering environment. Chemical engineering, and particularly process engineering, is not an exception, and there are many cases in this area which require optimization, usually in multivariable cases, with highly nonconvex equations representing the processes. Besides, models with both continuous and discrete variables, or involving algebraic and differential equations, can be found in such areas. Due to the complexity of the equations governing chemical processes, robust optimization methods are required to find the best designs in terms of the objectives of the designer. In the last years, stochastic optimization methods have been proposed and applied to the optimization of different systems, usually for the solution complex models the solution of complex models with a high number of variables and nonlineal functions. Such stochastic algorithms involve the analysis of the model for different regions on the solution space and selecting the best alternatives to conform the set of optimal solutions. These methods do not require calculation of derivatives unlike the traditional, deterministic methods; thus, stochastic methods are a good option to solve optimization problems for the complex process engineering models. This implies, of course, developing mathematical models to represent such systems. Nevertheless, modular process simulators are tools that can be helpful for optimization purposes, since they have great databases with the models of different process equipment and information for the physical properties of the components involved in many applications. Such simulators involve the use of black-box blocks, which contain the information of each component of the process. Nevertheless, in some cases, it is also possible to use equation-oriented approaches to represent user models inside the same simulation environment. Thus, the use of modular simulators combined with stochastic optimization methods is presented as a good alternative to obtain optimal designs for process equipment or even complete processes. In this book, some basic concepts about optimization and stochastic optimization are presented, and some techniques for the simultaneous design and optimization of process modules are discussed. Examples of application of those techniques to the optimal design of unit operations and chemical processes are also presented.

Acknowledgment

Juan Gabriel Segovia-Hernández I want to give special thanks to all the students who, over the years, have enriched these notes with their comments and contributions. I wish to make a very special recognition of the work and dedication for the preparation of this manuscript to Eduardo Sánchez-Ramírez, Juan José Quiroz-Ramírez, and César Ramírez-Marquez.

Fernando Israel Gómez-Castro I would like to acknowledge all my students for their comments in and outside the classroom. Each of your questions enhanced this work. Also, I acknowledge the contributions of my research group, particularly of Araceli Guadalupe Romero-Izquierdo and Mayra Margarita May-Vázquez. Finally, I appreciate the effort of Eduardo Sánchez-Ramírez, Juan José Quiroz-Ramírez, and César Ramírez-Marquez.

Editors

Juan Gabriel Segovia-Hernández is professor at the University of Guanajuato, Guanajuato, Mexico, since 2004, in the chemical engineering department. He has coedited one book and published over 110 papers in international journals, 8 book chapters, and several refereed conference proceedings. He was the national president of the Mexican Academy of Chemical Engineering (2013–2015). He is a member of the Mexican Academy of Sciences since 2012. His research interests include design, optimization, and control of intensified processes. He is currently a lecturer in several universities in Mexico and abroad. For more details on his research and publications, browse https://www.segovia-hernandez.com.

Fernando Israel Gómez-Castro is professor in the chemical engineering department at the University of Guanajuato, Guanajuato, Mexico, since 2012. He obtained the degree of ScD in chemical engineering in 2010, at the Institute of Technology of Celaya, Mexico. He is the author of 31 research papers published in national and international journals and 5 book chapters and reviewer of international journals such as *Chemical Engineering & Technology, Chemical Engineering Research and Design, Industrial & Chemistry Engineering Research,* and *Fuel,* among others. He is a member of the National Researchers System (Mexico) and the American Chemical Society. His biography has appeared in directories such as *Who's Who in the World* and *2000 Outstanding Intellectuals of the 21st Century.* He is currently a lecturer of subjects associated with mathematical optimization and its application to chemical process design, for both bachelor and graduate levels. Among his research interests can be mentioned the use of computational tools for the design and optimization of conventional and intensified chemical processes.

Contributors

Juan José Quiroz-Ramírez
Departamento de Ingeniería Química, División de Ciencias Naturales y Exactas, Campus Guanajuato, Universidad de Guanajuato

César Ramírez-Márquez
Departamento de Ingeniería Química, División de Ciencias Naturales y Exactas, Campus Guanajuato, Universidad de Guanajuato

Eduardo Sánchez-Ramírez
Departamento de Ingeniería Química, División de Ciencias Naturales y Exactas, Campus Guanajuato, Universidad de Guanajuato

1

Introduction to Optimization

1.1 What Is Optimization?

The term "optimization" may immediately lead us to a mathematical definition, given the background of any person related with science and/or engineering. Nevertheless, "optimization" can be defined from a more general point of view as "select the best alternative among a set of possibilities." This, certainly, implies that optimization procedures may occur for any person every day, even in an unconscious way. Moreover, since decisions are mainly taken by human beings, optimization should have definitely occurred from the beginning of mankind. It has been reported that one of the first registered optimization problems is the isoperimetric problem, which was solved by Queen Dido around 1000 BC, whereas the beginning of the systematic optimization procedures started with the solution of the brachistochrone problem, around 1694 (Diwekar, 2010). Nowadays, the size of optimization problems is considerably large, existing from considerably small scales to problems involving countries or even continents. The search of the best configuration for a polymer molecule may be performed through optimization algorithms (Venkatasubramanian et al., 1994). Finding the best alternative for production scheduling of a chemical plant is an optimization problem (Lin and Floudas, 2002). Moreover, selecting the best alternative for supply chain for the production of biofuels in a country (Leao et al., 2011) or an entire continent (Wetterlund et al., 2012) is also an optimization problem. As novel optimization problems are becoming more and more complex, the development of robust optimization algorithms is necessary. Such solution methods should be able to deal with a high number of continuous and discrete variables, nonconvex search spaces, multiple objectives, and other complexities shown by modern optimization problems.

1.2 Mathematical Modeling and Optimization

Although optimization may occur by trial-and-error procedures, such strategy can be quite expensive or even dangerous. In other cases, the number of possible solutions is considerably high; therefore, it is unpractical to test each of them. When those situations occur, rigorous optimization techniques are necessary. Moreover, such methods require, in several cases, counting with a mathematical model, which properly represents the phenomena or system of interest. A mathematical model is an abstract representation of the system under study, and it relates the important variables through mathematical expressions. Such mathematical equations can be expressed as equalities ($A = B$), inequalities ($A \leq B$ or $A \geq B$), or logical expressions ($A \rightarrow B$). Furthermore, relationships between the variables can be merely algebraic, which happens for static systems, or can be differential or integro-differential, which is observed in dynamic systems. Despite the type of mathematical equations and relationships conforms to the model, it should be used for better understanding the system under study, and obtaining information about the relationship between the different components of the system. Furthermore, the model will be beneficial for examining the effects of manipulating the input variables on the entire performance of the case of study. Moreover, it allows avoiding the high costs of multiple experiments and the risks of manipulating a not well understood system. Certainly, experimentation is necessary to obtain the unknown information required for the model or to validate the results obtained, but the required number of tests will be small.

An important concept, which is the first link between mathematical modeling and optimization, is the number of degrees of freedom. Let us assume a mathematical model with M independent equations and N variables. The number of degrees of freedom, F, is then defined as follows:

$$F = N - M \tag{1.1}$$

Thus, the degrees of freedom can be defined as a set of variables in excess, which avoids the model to be solved in a direct way. To solve the model, an **M×M** matrix should be obtained. Consequently, additional equations are required, which can be obtained by fixing F variables in a given value. Three situations can be observed when analyzing the number of degrees of freedom:

Case I. The number of equations is greater than the number of variables ($M > N$), and thus, the number of degrees of freedom is negative. This situation commonly implies that there are some errors in the model, and it is said that the problem is overspecified. Another possibility for the existence of this situation is that there are some dependent equations in the model, which should not be considered for computing M.

Case II. The number of equations is equal to the number of variables ($M = N$); thus, the number of degrees of freedom is zero. This implies that the system is an **M×M** matrix, and, if it consists of linear equations, there is only one solution to the problem. If the equations are nonlinear, multiple solutions can exist, but they are due to the roots of the nonlinear equations.

Case III. The number of variables is greater than the number of equations ($M > N$); thus, the number of degrees of freedom is positive. This is the case where we have variables whose values cannot be obtained using the model. Moreover, they should be assumed in a manner that the **M×M** matrix is completed and the model can be solved. Nevertheless, there is the problem of selecting proper values for those F variables. This implies that we can have a set of possible solutions, depending on the values of the degrees of freedom. As we mentioned previously, optimizing is selecting the best alternative from a set of possible solutions. Thus, the models with a positive value for the number of degrees of freedom can be optimized, and such problems are the focus of this book.

Example 1.1. The following set of equations is analyzed:

$$2x_1 + 3x_2 + 5x_3 + 6x_4 + x_5 = 10 \tag{1.2}$$

$$4x_2 + x_3 + 3x_4 + 4x_5 = 20 \tag{1.3}$$

$$x_2 + 2x_3 + x_4 + x_5 = 25 \tag{1.4}$$

$$3x_3 + x_4 + 2x_5 = 15 \tag{1.5}$$

$$2x_1 + 7x_2 + 6x_3 + 9x_4 + 5x_5 = 30 \tag{1.6}$$

This problem has been reported by Jiménez Gutiérrez (2003). The objective of this example is to compute the number of degrees of freedom for the system of Equations (1.2) through (1.6.) It can be observed that the number of equations, M, is 5, and the number of variables, N, is also 5. Subsequently, F would be 0 and the problem would have a single solution, as the equations are linear. Nevertheless, if the system of equations is observed in detail, it is clear that Equation (1.6) is the sum of Equations (1.2) and (1.3.) Thus, the equation is not independent and should not be considered for calculating the number of degrees of freedom. The truth is that the number of independent equations, M, is 4, and the number of degrees of freedom is equal to 1.

1.3 Classification of Optimization Problems

When a mathematical model is used for solving an optimization problem, it can be classified into different categories in terms of the number of degrees of freedom, including the type of mathematical relationships, equations, and variables. Depending on the type of optimization problem, the solution strategy will be different. In terms of the number of degrees of freedom, there are univariate problems, when there is only one degree of freedom; and multivariable problem, when there exist two or more degrees of freedom. For the univariate optimization problems, there are search methods, such as the golden section or the Fibonacci methods, which are considerably beneficial for solving that type of problems (Jiménez Gutiérrez, 2003). For multivariable optimization, more robust methods are required.

Optimization problems can also be classified in terms of the type of mathematical relationships on the model, which can be algebraic or differential/integro-differential. For both cases, uncertainties may or may not occur for the model components. If the model has only algebraic equations and there are no uncertainties, we discuss about a classical mathematical programming problem. If there are uncertainties, the case is known as a stochastic programming problem. When the model consists of differential/integro-differential relationships, but there are no uncertainties, we discuss about an optimal control problem. Finally, if there are uncertainties, a stochastic optimal control problem arises. Figure 1.1 shows this classification in a graphical way.

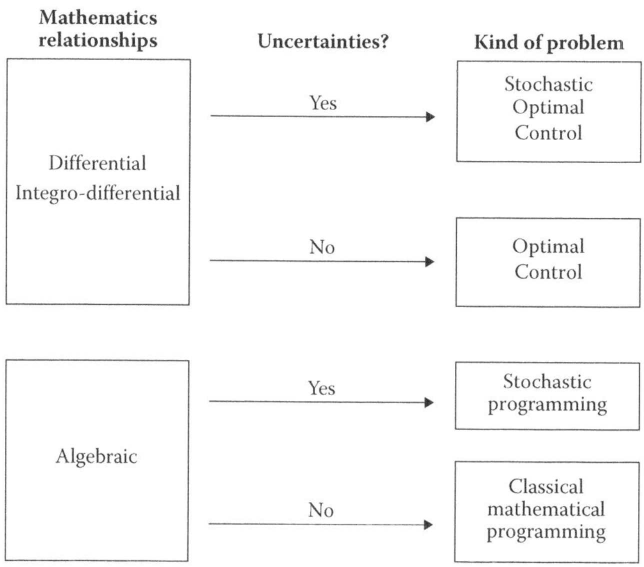

FIGURE 1.1
Classification of the optimization problems in terms of the type of mathematical relationships.

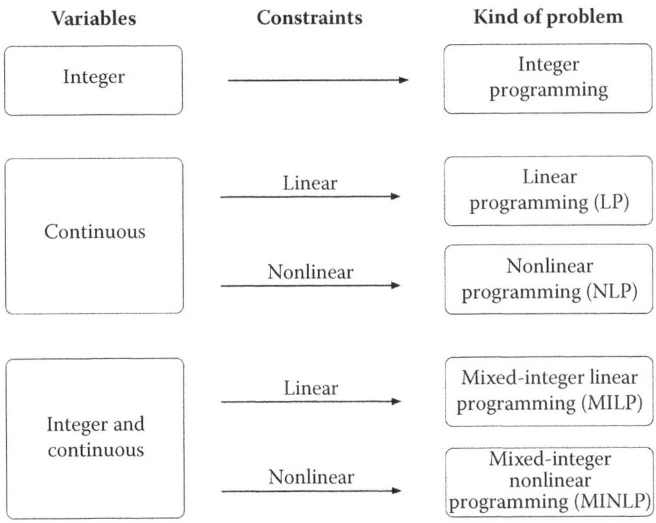

FIGURE 1.2
Classification of the optimization problems in terms of the type of variables and constraints.

Optimization problems can also be classified in terms of the type of variables and the type of equations in the model. A problem may consist only of integer variables, continuous variables, or a combination of continuous and integer variables. When we have a problem with integer variables, it is called an integer optimization problem. For a system with continuous variables, we may have linear or nonlinear constraints. If the constraints are linear, there is a linear programming (LP) problem. If the constraints are nonlinear, it is a nonlinear programming (NLP) problem. On the other hand, if the problem involves both integer and continuous variables, and the constraints are linear, the problem is a mixed-integer linear programming (MILP) problem. If the constraints are nonlinear, we have a mixed-integer nonlinear programming (MINLP) problem. This classification can be better observed in Figure 1.2.

1.4 Objective Function

We have mentioned that optimizing implies selecting the best alternative among a set of possibilities. Nevertheless, the term "the best" is quite relative, and the selection of the best alternative strongly depends on the personal opinion of the decision maker. Thus, to avoid taking subjective decisions, a more trustworthy, numeric criteria should be established, which allows selecting the solution independent to the personal criteria of the one responsible of taking the decision. That criteria is called the objective function.

The objective function can be defined as a way for measuring the effectiveness of the system (Sarker and Newton, 2008), or a way for measuring the performance of the system (Pierre, 1986). In other words, it indicates whether a given solution is good in comparison to others, or if it can be considered as the best solution. In Figure 1.3a, a one-variable objective function is shown. In Figure 1.3a, the optimal solution is the one marked as x^*. For that solution, the objective function takes a value of $f(x^*)$. It can be observed that there is no other value of $f(x)$ smaller than $f(x^*)$ for any other x. Thus, it is said that the solution is a global minimum. In the case of the objective function in Figure 1.3a, finding the optimal solution is quite simple, implying the use of the first derivative criteria. Nevertheless, when the number of decision variables is higher, the solution of the optimization problem is not that easy. Figure 1.3b shows an objective function with two independent variables. It can be observed that there are two points, which can be classified as minimums, $\bar{x}_1^*$ and $\bar{x}_2^*$. For both solutions, the gradient is equal to zero; thus, they are both optimal solutions. Nevertheless, the value of the function evaluated for $\bar{x}_2^*$ is lower than the value of the function for $\bar{x}_1^*$. Moreover, $f(\bar{x}_2^*)$ is the lowest value the function can consider, and it is a global minimum. The solution given by $f(\bar{x}_1^*)$ is a minimum, but it is the lowest value of the function only for the surroundings of $\bar{x}_1^*$. Thus, it is known as a local minimum.

An unconstrained optimization problem can be stated in a general form as follows:

$$\text{optimize } Z = z(\bar{x}) \tag{1.7}$$

where the term *"optimize"* is replaced by *"min"* for minimization and *"max"* for maximization, depending on what type of solution is desired. Z is a

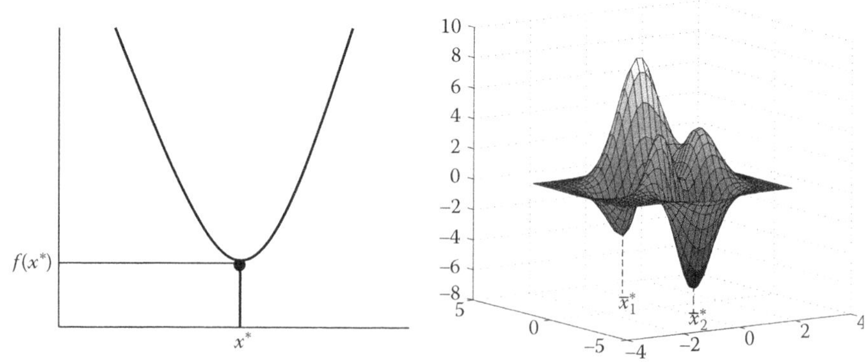

(a) (b)

FIGURE 1.3
Objective functions. (a) Objective function for a single variable and (b) objective function for two variables.

variable assigned to the objective function, and $z(\bar{x})$ is the objective function, expressed in terms of the vector of decision variables $\bar{x}$.

1.5 Optimization with Constraints: Feasible Region

The objective function is the heart of an optimization problem, since it is the function that should be minimized or maximized. When the optimization situation involves only the analysis of the objective function, as shown in Figure 1.3, it is said that the problem is unconstrained, and the search for the solution occurs in the entire set of R^n. Nevertheless, for most of the practical problems, the variables are not unbounded; thus, the search space is reduced by the introduction of bounds for the variables. Furthermore, most of the variables appearing on the objective function are related to other variables through a mathematical model, which implies that the search space for the objective function is even more reduced. The region in the n-dimensional space limited by those bounds for the variables and the equations relating the set of variables is known as the feasible region. In Figure 1.4a, a feasible region in a bidimensional space can be observed, which is enclosed by the constraints $h_1(x)$, $h_2(x)$, and $g_1(x)$. On the other hand, Figure 1.4b shows a tridimensional feasible region, where the constraint $g(\bar{x})$ is a plane, and the objective function is a surface with various minimums. Until now, two types of constraints have been mentioned, $h(\bar{x})$ and $g(\bar{x})$. The difference between the two types of constraints is explained. The equations $h(\bar{x})$ are equality constraints, i.e., they have the structure $h(\bar{x}) = 0$. Such constraints should be strictly complied. Thus, they reduce even more the number of feasible solutions. On the other hand, the equations $g(\bar{x})$ are inequality constraints, i.e.,

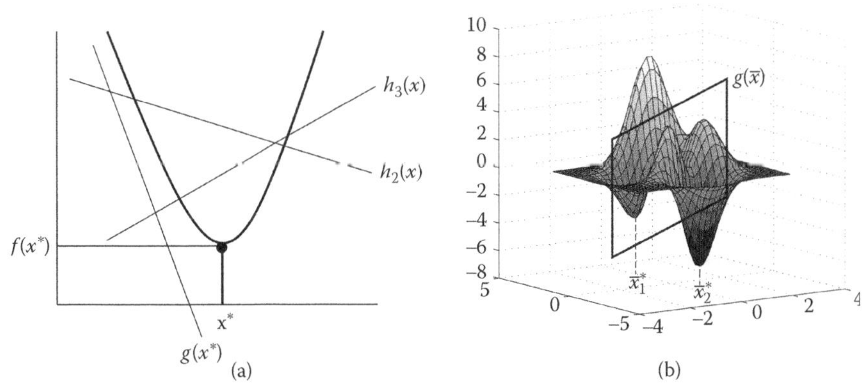

FIGURE 1.4
Constrained objective functions. (a) Feasible region in two dimensions and (b) feasible region in three dimensions.

$g(\overline{x}) \leq 0$. Such constraints can be complied as an equality, $g(\overline{x}) = 0$, for which it is said that the constraint is active; or they can be complied as an inequality, $g(\overline{x}) < 0$, for which it said that the constraint is inactive. The inequality constraints, thus, are more easily complied, because they allow the feasible region to include higher number of solutions.

In Figure 1.4a, it can be observed that the objective function and the optimal solution are the same that the one shown in Figure 1.3a, since it complies with both equality constraints. If the inequality constraint implies that the feasible region is located to its right side, then the solution also complies with the inequality constraint. Nevertheless, the optimal solution in the constrained problem is not the same as for the unconstrained problem in all the cases. The unconstrained solution could be outside the feasible region of the constrained problem; thus, the solution for the last case may be different than the solution for the first one. The optimal solution on a constrained problem should simultaneously comply with the objective function and the entire set of constraints. In Figure 1.4b, the feasible region depends on the form of the constraint, i.e., $g(\overline{x}) \leq 0$ or $g(\overline{x}) \geq 0$. Thus, the feasible region may include only the minimum given by $\overline{x}_1^*$, or the other one, given by $\overline{x}_2^*$. Since the solution should comply with the inequality constraint, only one of the minimal points could be achieved with the given $g(\overline{x})$, and the constraint may or may not avoid reaching the global minimum of the unconstrained objective function.

A constrained optimization problem can be represented in a general way as follows:

$$\begin{aligned} &\text{optimize } Z = z(\overline{x}) \\ &\text{s.t.} \\ &h(\overline{x}) = 0 \\ &g(\overline{x}) \leq 0 \end{aligned} \tag{1.8}$$

As aforementioned, $h(\overline{x})$ are equality constraints and $g(\overline{x})$ are inequality constraints, which conform the feasible region for the constrained optimization problem.

1.6 Multiobjective Optimization

At this point, the objective function has been mentioned as a measurement of the goodness of a given solution. It has been stated that an optimization problem, where there is only the necessity of maximizing/minimizing a given objective function, is known as an unconstrained problem. On the other hand, the optimization problem may involve constraints, which limits the solution space for the objective function. Nevertheless, there are cases on which it is desired to simultaneously optimize two or more objective functions subject to the same set of constraints and variables. This is called multiobjective optimization. This type of problems can be represented in the following general formulation:

$$\begin{aligned}
&\text{optimize } \bar{Z} = (Z_1, Z_2, \ldots, Z_k) \\
&\text{s.t.} \\
&h(\bar{x}) = 0 \\
&g(\bar{x}) \leq 0
\end{aligned} \tag{1.9}$$

where k is the total number of objectives, $\bar{Z}$ is the vector of objective functions, and $Z_1 = z_1(\bar{x})$, $Z_2 = z_2(\bar{x})$, and so on are the individual objective functions. As it is observed, the feasible regions, i.e., the constraints of the optimization problem, are the same for all the objectives. Thus, the only difference with the optimization problem given in Equation 1.8 is that there are more than one objective functions, and the solution complies with all of them and also with the constraints.

Let us imagine about a problem with two objectives. Both objective functions may have their minimum (or maximum) on the same (or almost on the same) point $\bar{x}$, as shown in Figure 1.5a. Nevertheless, such a problem could even be stated as a problem with a single objective. A more interesting situation occurs when the minimum of the objective functions occur for a different set of $\bar{x}$ values, which implies that when one of the objectives is being reduced, the other one increases (Figure 1.5b). When two objective functions follow such performance, it is mentioned that they are in competence. This implies that, when the minimum for one of the objective functions occurs, the other objective has a value far from its own minimum. Thus, assuming, in general, a situation with k objective functions, the solution to the optimization problem should be a compromise

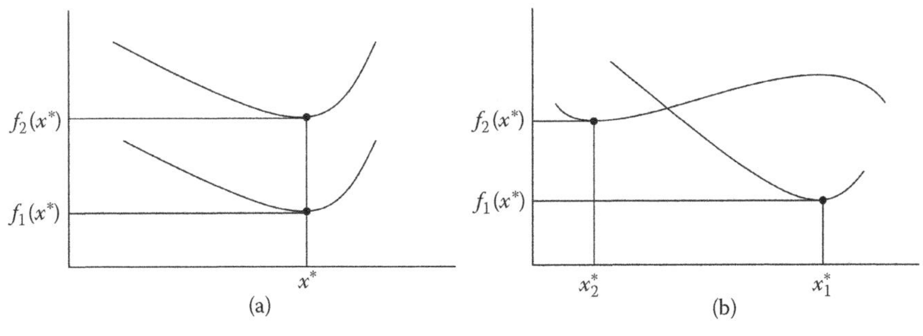

FIGURE 1.5
(a) Objective functions not competing and (b) objective functions in competence.

among all the objectives, since the better solution for one of the objectives could be a considerably inadequate solution for the others. An important concept in multiobjective optimization is the nondominated solution. In mathematical terms, it can be mentioned that $\overline{x}^*$ is a nondominated solution for a minimization problem if, for any other feasible solution $\overline{x}$, the following relationship is true:

$$Z_p(\overline{x}) \leq Z_p(\overline{x}^*) \tag{1.10}$$

where $p = 1, 2, \ldots, k$, and Z_p is a given objective function.

This implies that a nondominated solution is better than any other solution $\overline{x}$ for at least one of the objective functions. Certainly, there could be several nondominated solutions depending on the nature and number of the objective functions. For a maximization problem, $\overline{x}^*$ is a nondominated solution if:

$$Z_p(\overline{x}) \geq Z_p(\overline{x}^*) \tag{1.11}$$

The entire set of nondominated solutions is known as the Pareto front, in memory of the economist and sociologist Vilfredo Pareto. The Pareto front describes the different optimal solutions obtained from the multiobjective optimization procedure. An example of Pareto front for two objective functions, Z_1 and Z_2, is presented in Figure 1.6. Each point represents a nondominated solution; thus, each one is an optimal solution. It can be observed that, for small values of Z_1, Z_2 tends to have high values, and vice versa: when Z_2 decreases, Z_1 is increased. Thus, the objective functions are in competence. The extreme points are the optimal solutions when considering only one of the objective functions in the optimization procedure. The last point to the

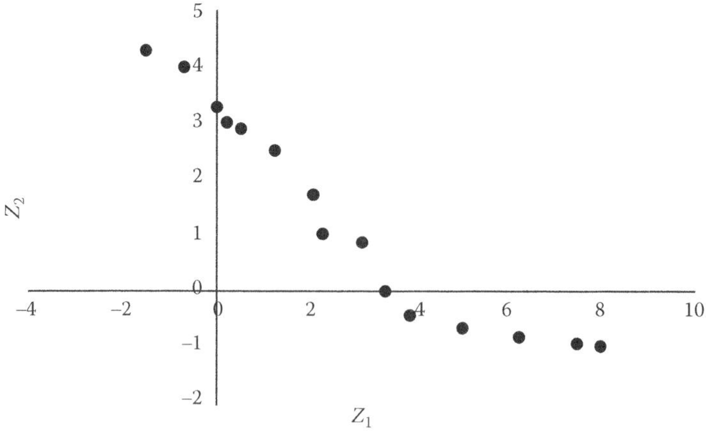

FIGURE 1.6
A simplified Pareto front.

left indicates the minimum for Z_1, where Z_2 has a maximum. On the other hand, the last point to the right is the minimum for Z_2, and Z_1 shows a maximum on that solution.

To solve a multiobjective optimization problem, any traditional optimization methodology can be used. Nevertheless, the multiobjective problem is modified for obtaining a single objective function which includes the effect of all the individual objectives. There are different strategies to generate such a function, which can be classified as follows:

- Generating methods: In these methods, the goal is to generate an approximation to the Pareto front by solving multiple optimization problems.
- Preference-based methods: In these methods, the goal is to look for the solution that is as close as possible to the solution desired by the decision maker.

In this section, two generating methods are discussed, since those methods allow us to visualize an entire set of potential solutions to the multiobjective problem.

1.6.1 Weighted Sum Method

In this method, the optimization problem shown in Equation 1.9 is modified as follows:

$$\begin{aligned} &\text{optimize } Z_{\text{mult}} = \sum_{i=1}^{k} \omega_i Z_i \\ &\text{s.t.} \\ &h(\bar{x}) = 0 \\ &g(\bar{x}) \leq 0 \end{aligned} \tag{1.12}$$

Thus, the vector of objectives is substituted by a linear combination of the individual objectives. The parameters ω_i are known as weights. In the weighing method, the function Z_{mult} is used for generating the Pareto front. In the first approach, k individual optimization problems are solved for each objective function. In other words, the problem is first solved by setting one of the ω_i's as 1 and the other weights as zero, and so on, until the optimal for each individual objective function has been found. Those points represent the extremes at the Pareto front. Then, the intermediate solutions are obtained by testing different combinations of ω_i, which provide more or less importance to each Z_i. Certainly, particular care should be provided on the scale of each objective functions for a proper selection of the values of the weights, in order to have equality on the contributions of each individual objective to Z_{mult}.

1.6.2 Constraint Method

In the constraint method, the optimization problem represented in Equation 1.9 is modified as follows:

$$\begin{aligned}
&\text{optimize } Z_i \\
&\text{s.t.} \\
&Z_j \leq \varepsilon_j \quad (j = 1, 2, \ldots, k; j \neq i) \\
&h(\bar{x}) = 0 \\
&g(\bar{x}) \leq 0
\end{aligned} \tag{1.13}$$

Thus, one of the objective functions is selected to be optimized, and the others are moved to the set of constraints as inequality constraints, with a right-side term ε_j. As a first step, individual optimization problems are solved for the objectives Z_j, obtaining the lower and upper limits for each of those objectives. Then, using the range of values for the individual objectives, the problem is discretized and a given number of internal points are selected. The selection of the data can be performed through a sampling methodology. Each point will be a set of values for the ε_j terms. Thus, for n selected points, n optimization problems should be solved, and each solution will be a nondominated one.

For a more detailed discussion of the presented multiobjective optimization methods, the reader is referred to the study of Diwekar (2010). Other multiobjective optimization methods are consulted in the study of Marler and Arora (2004).

1.7 Process Optimization

In general, engineering systems are typically good candidates for optimization, owing to its high number of degrees of freedom. Furthermore, the equations modeling such systems are typically nonlinear and may involve both algebraic and differential relationships. Thus, a rigorous approach to solve optimization problems in engineering is mandatory. Chemical engineering, and in particular, process engineering is not an exception. Imagining the heart of a chemical process, the reactor, such systems are typically modeled by nonlinear equations, including, as an example, the Arrhenius equation to represent the changes on the kinetic constant with temperature. Separation systems are also represented by models with a high number of nonlinear equations, e.g., the thermodynamic relationships modeling phase equilibrium. Furthermore, when the unit operations consist of various separation stages, the number of equations is increased. The number of variables in the models of chemical processes and the number of degrees of freedom are high. Variables such as temperature, pressure, mass of catalyst, holdup, coolant flow rate, among others are typically degrees of freedom in process engineering. In the case of the objective functions, there are some typical objectives in process engineering, such as minimizing the total annual costs, maximizing the profit, minimizing the environmental impact, maximizing the social impact, and minimizing the control effort. Constraints are typically provided by the mathematical models itself, but also by the inherent characteristics of the process, e.g., lower and upper limits for the variables. Lower limits are typically provided by the positive nature of most of the physical variables; whereas, upper limits are provided by operational limitations. When designing equipment, constraints may also occur owing to limitations on available space for installation. From the last lines, it can be deduced that process engineering is a source of various optimization problems, where most of them are multivariable ones. Thus, robust strategies are required for solving such situations, considering the high number of degrees of freedom that can be implied. In general terms, process optimization problems can be solved by using three approaches: mathematical programming (Grossmann et al., 1999; Caballero and Grossmann, 2004), stochastic optimization methods (Androulakis and Venkatasubramanian, 1991; Cardoso et al., 2000), and hybrid methods (Banga et al., 2003; Mohammadhasani Khorasany and Fesanghary, 2009). The first approach is the most rigorous approach, since it solves the entire model and finds the optimum solution through the Calculus principles. In the second approach, the solution is searched in the entire feasible region, using some criteria to reach the global optimum. In the third approach, a stochastic method is initially used to reach a zone close to the global optimum, and then a mathematical programming method is used to ensure reaching the global optimum.

In the modern process optimization methods, the entire mathematical models for different process equipment contained in the process simulators are linked with optimization software in order to use formal strategies and reach the best designs in terms of given objective functions. This approach has the advantage of null or low necessity of programming the model for each equipment or the thermodynamic/kinetic relationships. Thus, the efforts are focused on developing proper objective functions and finding the best optimization strategy for those models. This is particularly helpful for complex systems, with recycles, energy or mass integration, formation of multiple phases, and so on. In this book, some insights of mathematical programming are discussed, but the main focus is to describe the combination of stochastic optimization tools and process simulators (in particular, Aspen Plus®) to solve different types of optimization problems in process engineering. Study cases for the single-objective and multiobjective optimization of isolated process equipment and entire processes are presented, and the methodology for the solution of each case is described in detail.

References

I.P. Androulakis, V. Venkatasubramanian, 1991, A genetic algorithmic framework for process design and optimization, *Comput. Chem. Eng.*, 15(4), 217–228.

J.R. Banga, E. Balsa-Canto, C.G. Moles, A.A. Alonso, 2003, Improving food processing using modern optimization methods, *Trends Food Sci. Technol.*, 14(4), 131–144.

J.A. Caballero, I.E. Grossmann, 2004, Design of distillation sequences: From conventional to fully thermally coupled distillation systems, *Comput. Chem. Eng.*, 28(11), 2307–2329.

M.F. Cardoso, R.L. Salcedo, S. Feyo de Azevedo, D. Barbosa, 2000, Optimization of reactive distillation processes with simulated annealing, *Chem. Eng. Sci.*, 55(21), 5059–5078.

U. Diwekar, 2010, *Introduction to Applied Optimization*, 2nd edition, New York, NY: Springer.

I.E. Grossmann, J.A. Caballero, H. Yeomans, 1999, Mathematical programming approaches to the synthesis of chemical process systems, *Korean J. Chem. Eng.*, 16(4), 407–426.

A. Jiménez Gutiérrez, 2003, *Diseño de Procesos en Ingeniería Química*, Barcelona: Reverté.

R.R.C.C. Leao, S. Hamacher, F. Oliveira, 2011, Optimization of biodiesel supply chains based on small farmers: A case study in Brazil, *Bioresour. Technol.*, 102(19), 8958–8963.

X. Lin, C.A. Floudas, 2002, Continuous-time optimization approach for medium-range production scheduling of a multiproduct batch plant, *Ind. Eng. Chem. Res.*, 41(16), 3884–3906.

R.T. Marler, J.S. Arora, 2004, Survey of multi-objective optimization methods for engineering, *Struct. Multidiscipl. Optim.*, 26(6), 369–395.

R. Mohammadhasani Khorasany, M. Fesanghary, 2009, A novel approach for synthesis of cost-optimal heat exchanger networks, *Comput. Chem. Eng.*, 33(8), 1363–1370.

D.A. Pierre, 1986, *Optimization Theory with Applications*, New York, NY: Dover.

R.A. Sarker, C.S. Newton, 2008, *Optimization Modelling: A Practical Approach*, Boca Raton, FL: CRC Press.

V. Venkatasubramanian, K. Chan, J.M. Caruthers, 1994, Computed-aided molecular design using genetic algorithms, *Comput. Chem. Eng.*, 18(9), 833–844.

E. Wetterlund, S. Leduc, E. Dotzauer, G. Kindermann, 2012, Optimal localization of biofuel production on a European scale, *Energy*, 41(1), 462–472.

2

Deterministic Optimization

2.1 Introduction

Deterministic optimization, or mathematical programming, is the classical way to perform optimization through mathematical models. It is also the most rigorous way to obtain the best solution for a given problem. Mathematical programming methods are based on the principles of calculus; thus, the solution of optimization problems basically involves finding stationary points where the gradient vector is equal to zero. This implies that the solution depends strongly on the initial values and on the convexity of the objective function and the constraints. Depending on the type of optimization problem, the solution methods can vary. Stochastic programming involves the use of probabilistic distributions to approach the variables or functions presenting uncertainties. Optimal control problems can be solved by using calculus of variations or by using discretization methods. In the case of classical mathematical programming, different approaches can be used, depending on the nature of the problem. In this chapter, the solution of nonlinear programming (NLP) problems is discussed to illustrate the main characteristics of the mathematical programming methods for the solution of optimization problems. For a more detailed discussion of other deterministic optimization methods, the reader is referred to the works of Pierre (1986), Diwekar (2010), Biegler (2010), Liberzon (2012), and Shapiro et al. (2014).

2.2 Single-Variable Deterministic Optimization

We start with the simplest type of mathematic optimization, which is an unconstrained, one-variable optimization problem. This is helpful to better understand the solution methods of multivariable problems, with and without constraints. The well-known first derivative criterion is that, if we have a given function f(x), we can find a stationary point x^* if we derive the function and equal the derivative to zero, and then solve for x. That stationary point could be an optimal (minimum or maximum) or not, depending on the characteristics of the function. If the function is convex for any x (the second derivative is positive or zero), then x^* is a minimum. If the function is concave for any x (the second derivative is negative), then x^* is a minimum.

Nevertheless, if the function is neither concave nor convex, then the stationary point is neither a maximum nor a minimum, but a saddle point. Some examples are discussed in this chapter before dealing with multivariable optimization.

Example 2.1: Find the minimum value for the function $f(x) = x^4 + 2$.

The function is shown in Figure 2.1. The first derivative of the function can be expressed as follows:

$$f'(x) = 4x^3 \tag{2.1}$$

If the derivative is equal to zero, the solution is $x^* = 0$. The second derivative of the function is given by

$$f''(x) = 12x^2 \tag{2.2}$$

If we evaluate the second derivative, it can be observed that it remains positive for any x, except for x = 0, where $f''(0) = 0$. Thus, the obtained solution is indeed a minimum. This can be observed in Figure 2.1, where there is a single minimum and the function is always convex.

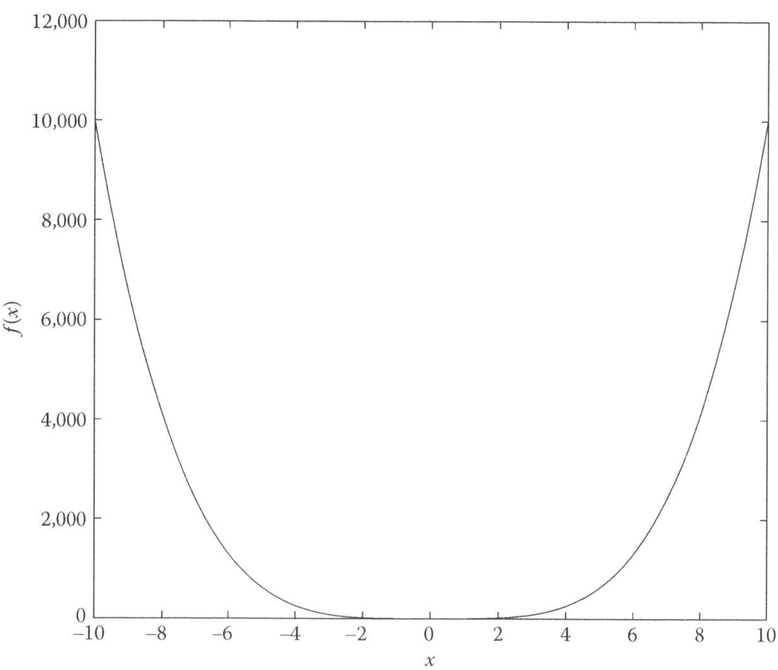

FIGURE 2.1
Function $f(x) = x^4 + 2$.

Example 2.2: Find the minimum value for the function f(x) = x^3.

The function is shown in Figure 2.2. The first derivative of the function can be expressed as follows:

$$f'(x) = 3x^2 \quad (2.3)$$

If the derivative is equal to zero, a quadratic equation can be obtained. The roots of the equation are $x_1 = x_2 = 0$. Thus, an optimal solution can be expected. Nevertheless, we obtain the second derivative of the function as follows:

$$f''(x) = 6x \quad (2.4)$$

When evaluating the second derivative, it can be observed that it is positive for $x > 0$, but it is negative for $x < 0$. Thus, the function is not convex, and the solutions are not optimal points. For $x = 0$, the first derivative is zero, but it remains positive for any x. Thus, the solution represents a saddle point, as observed in Figure 2.2.

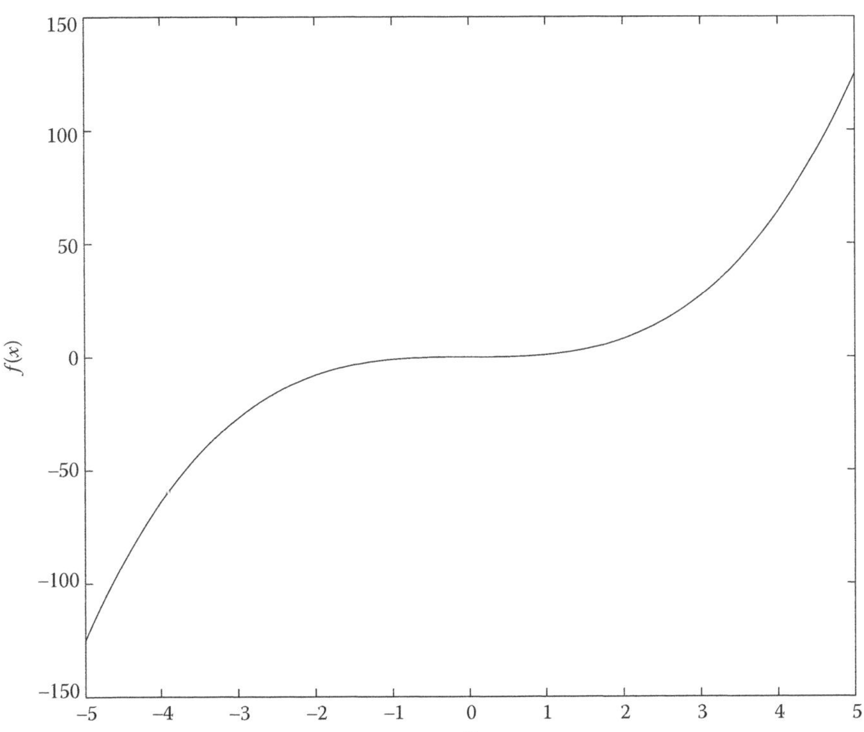

FIGURE 2.2
Function $f(x) = x^3$.

Example 2.3: Find the minimum value for the function $f(x) = 2x^3 + 3x^2 - 12x + 1$.

The function is shown in Figure 2.3. The first derivative is expressed as follows:

$$f'(x) = 6x^2 + 6x - 12 \tag{2.5}$$

If the derivative is equal to zero, a quadratic equation is obtained. The roots of the equation are $x_1 = -2$ and $x_2 = 1$. Thus, two optimal solutions can be expected. Nevertheless, we obtain the second derivative of the function as follows:

$$f''(x) = 12x + 6 \tag{2.6}$$

When evaluating the second derivative, it can be observed that it is positive for $x > -1/2$, but it is negative for $x < -1/2$. Therefore, if we look at the global picture, the function is not convex. Nevertheless, a further analysis of the first derivative indicates changes in its sign in the surroundings of x_1 and x_2. This implies that the solutions are indeed optimal points, but there can be many of them in the entire feasible region. In Figure 2.3, it can be observed that x_1 is a maximum whereas x_2 is a minimum.

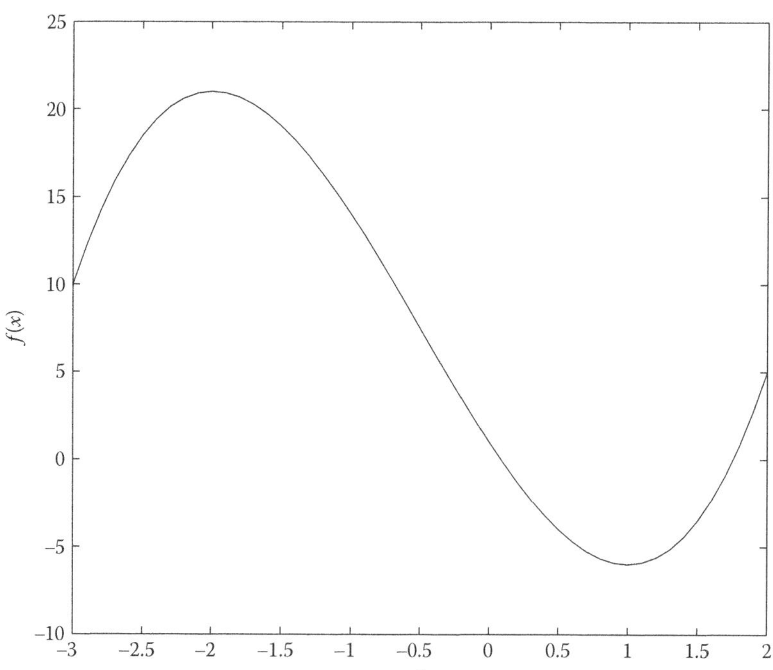

FIGURE 2.3
Function $f(x) = 2x^3 + 3x^2 - 12x + 1$.

2.3 Continuity and Convexity

For multivariable optimization, two important concepts are continuity and convexity. The functions to be optimized are desired to have those properties, although, even if they are neither continuous nor convex, the functions can be optimized with certain limitations. In this section, continuity and convexity of functions are described, and the importance of such properties on optimization is stressed.

A given function $f(\bar{x})$ is continuous in a point $\bar{x}_0$ if the following equality is true:

$$f(\bar{x}_0) = \lim_{\bar{x} \to \bar{x}_0} f(\bar{x}) \tag{2.7}$$

If Equation 2.7 is true for any value of $\bar{x}$ in the domain of the function, where $\bar{x} \in R^n$, then the function is continuous in the entire domain. An example of a continuous function is shown in Figure 2.4. It can be observed that the function is defined for any value of $\bar{x}$, and it does not exhibit any disruption. The limits for the function can be evaluated for any $\bar{x}$, and they are equal to the value of the function. Thus, it is continuous.

A noncontinuous function is presented in Figure 2.5. The function is defined for almost any value of $\bar{x}$. Nevertheless, when $\bar{x}$ is close to the point

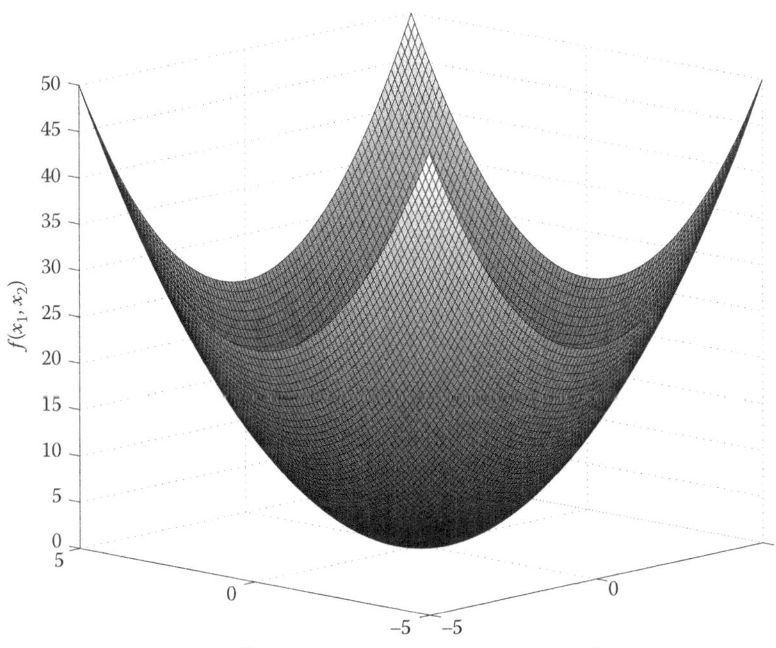

FIGURE 2.4
Continuous function.

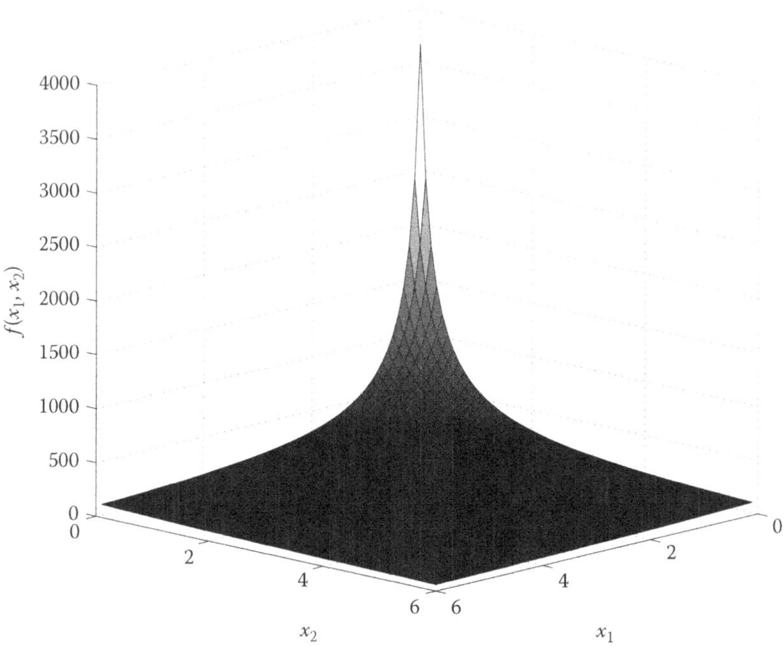

FIGURE 2.5
Noncontinuous function.

$\bar{x}_0 = [0\ 0]^T$, the function grows, and it will reach infinity at the point $\bar{x}_0$. Thus, the function is not defined at $\bar{x}_0$, and it is noncontinuous.

Because most of the deterministic optimization methods are based on the calculation of derivatives, dealing with continuous functions ensures that the derivatives exist for all feasible regions. For noncontinuous functions, if the solution is close to a discontinuity point, the derivative will not exist and problems will arise with the optimization algorithm. Nevertheless, through a proper analysis of the functions and a good selection of the limits of the variables, it is possible to avoid the discontinuities in several cases.

Other desired property of deterministic optimization is convexity. This is because, if the objective function is convex, it can be ensured that the obtained solution for the optimization problem is a global minimum. On the other hand, if the objective function is concave, the solution will lead to a global maximum. If the function is neither concave nor convex, it may have multiple local optimums, or may not have an optimum at all. In Figure 2.6, examples of convex and nonconvex functions are shown.

To determine the convexity of a given multivariable function, the Hessian matrix must be obtained. This matrix is named after the German mathematician Ludwig Otto Hesse (1811–1874) and contains information about all the second derivatives of a function. For a given function f $(\bar{x})$, the Hessian

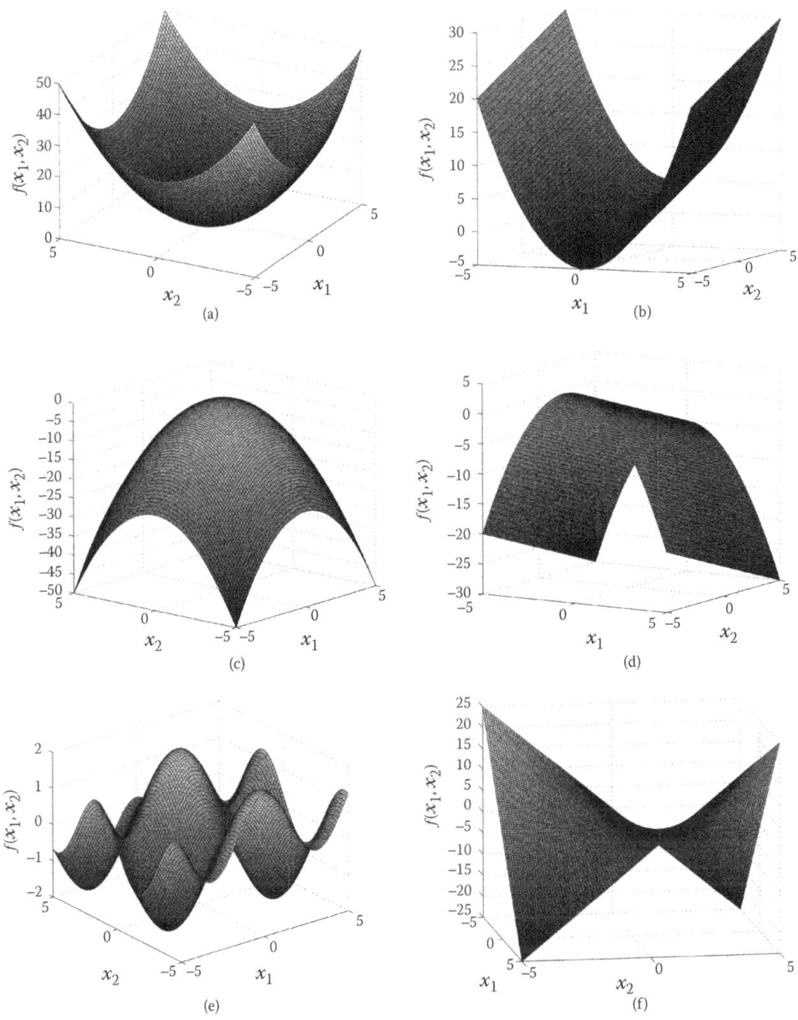

FIGURE 2.6
Convex and nonconvex functions: (a) strictly convex function, (b) convex function, (c) strictly concave function, (d) concave function, (e) nonconvex function with various local optima, and (f) nonconvex function with no optimal solution.

matrix, $\overline{\overline{H}}\,[\mathrm{f}\,(\overline{x})]$, or, in a more simplified representation, $\overline{\overline{H}}\,(\overline{x})$, is expressed as follows:

$$\overline{\overline{H}}\,(\overline{x}) = \begin{bmatrix} \dfrac{\partial^2 \mathrm{f}}{\partial x_1^2} & \cdots & \dfrac{\partial^2 \mathrm{f}}{\partial x_1\,\partial x_n} \\ \vdots & \ddots & \vdots \\ \dfrac{\partial^2 \mathrm{f}}{\partial x_n\,\partial x_1} & \cdots & \dfrac{\partial^2 \mathrm{f}}{\partial x_n^2} \end{bmatrix} \tag{2.8}$$

TABLE 2.1

Convexity/Concavity Criteria for a Function

Characteristic Values (λ_i^{EIG})	Nature of the Hessian Matrix	Nature of the Function
All positive	Positive definite	Strictly convex
All positive or zero	Positive semidefinite	Convex
All negative	Negative definite	Strictly concave
All negative or zero	Negative semidefinite	Concave
Positive, negative, and zero	Not definite	Neither convex nor concave

The convexity (or nonconvexity) of a function can be determined in terms of the characteristic values of its Hessian matrix. In Table 2.1, the criteria for convexity are presented. To obtain a global minimum, all the characteristic values of the Hessian matrix must be positive. Similarly, to obtain a global maximum, λ_i^{EIG}, all the characteristics values of the Hessian matrix, must be negative. Some cases of every type of function are discussed in Table 2.1.

Example 2.4: Determine if the following functions are convex or concave:

a. $f(x_1, x_2) = x_1^2 + x_2^2$
b. $f(x_1, x_2) = x_1^2 + x_2$
c. $f(x_1, x_2) = -x_1^2 - x_2^2$
d. $f(x_1, x_2) = -x_1^2 - x_2$
e. $f(x_1, x_2) = \cos(x_1) + \sin(x_2)$
f. $f(x_1, x_2) = x_1 x_2$

a. The Hessian matrix of the function is expressed as follows:

$$\overline{\overline{H}}(\overline{x}) = \begin{bmatrix} 2 & 0 \\ 0 & 2 \end{bmatrix} \tag{2.9}$$

The characteristic values of the Hessian matrix are $\lambda_1^{EIG} = 2$ and $\lambda_2^{EIG} = 2$. Both the values are positive and constant. Thus, the function is strictly convex, as can be observed in Figure 2.6a.

b. The Hessian matrix of the function is expressed as follows:

$$\overline{\overline{H}}(\overline{x}) = \begin{bmatrix} 2 & 0 \\ 0 & 0 \end{bmatrix} \tag{2.10}$$

The characteristic values of the Hessian matrix are $\lambda_1^{EIG} = 0$ and $\lambda_2^{EIG} = 2$. A characteristic value is positive, but the other one is zero. Thus, the function is convex. A representation of the function can be observed in Figure 2.6b.

c. The Hessian matrix of the function is expressed as follows:

$$\overline{\overline{H}}(\bar{x}) = \begin{bmatrix} -2 & 0 \\ 0 & -2 \end{bmatrix} \tag{2.11}$$

The characteristic values of the Hessian matrix are $\lambda_1^{EIG} = -2$ and $\lambda_2^{EIG} = -2$. Both the characteristic values are negative. Thus, the function is strictly concave, as can be observed in Figure 2.6c.

d. The Hessian matrix of the function is expressed as follows:

$$\overline{\overline{H}}(\bar{x}) = \begin{bmatrix} -2 & 0 \\ 0 & 0 \end{bmatrix} \tag{2.12}$$

The characteristic values of the Hessian matrix are $\lambda_1^{EIG} = 0$ and $\lambda_2^{EIG} = -2$. A characteristic value is negative, but the other one is zero. Thus, the function is concave, as can be observed in Figure 2.6d.

e. The Hessian matrix of the function is expressed as follows:

$$\overline{\overline{H}}(\bar{x}) = \begin{bmatrix} -\cos(x_1) & 0 \\ 0 & -\sin(x_2) \end{bmatrix} \tag{2.13}$$

The characteristic values of the Hessian matrix are $\lambda_1^{EIG} = -\cos(x_1)$ and $\lambda_2^{EIG} = -\sin(x_2)$. In this case, the characteristic values depend on the values taken by the variables x_1 and x_2. If x_1 is on the first or the fourth quadrant, λ_1^{EIG} is negative. On the other hand, if x_1 is on the second or third quadrant, λ_1^{EIG} is positive. Similarly, if x_2 is on the first or the second quadrant, λ_2^{EIG} is negative. Finally, if x_2 is on the third or the fourth quadrant, λ_2^{EIG} is positive. Thus, the function is convex for some intervals and concave for others. However, in general, the function is not convex and may show multiple minima or maxima. This can be observed in Figure 2.6e.

f. The Hessian matrix for this function is expressed as follows:

$$\overline{\overline{H}}(\bar{x}) = \begin{bmatrix} 0 & 1 \\ 1 & 0 \end{bmatrix} \tag{2.14}$$

The characteristic values of the Hessian matrix are $\lambda_1^{EIG} = -1$ and $\lambda_2^{EIG} = 1$. A characteristic value is negative and the other is positive. Thus, the Hessian matrix is undefined, and the function is neither convex nor concave. Unlike the function discussed in problem (e), the characteristic values for this example will not change with variations on $\bar{x}$, and the function will always remain nonconvex. The function is shown in Figure 2.6f. It can be observed that there is neither a well-defined minimum nor a maximum. Thus, if the function is optimized using a gradient-based approach, the solution will rely on a saddle point.

2.4 Unconstrained Optimization

The simplest case of deterministic, nonlinear optimization occurs when the problem has no constraints, i.e., the objective function must be optimized for any $\overline{x} \in R^n$ on its domain. For linear programming, constraints must always exist, because a linear function continues increasing (or decreasing) its value when the decision variables change; thus, no optimal solution can be obtained for a linear objective function without constraints. For nonlinear optimization, most of the solution methods are based on the calculation of derivatives to perform a search for stationary points. A given point $\overline{x}^*$ is a stationary point of the function f $(\overline{x})$ if it complies with the following condition:

$$\nabla f\,(\overline{x}^*) = 0 \tag{2.15}$$

Equation 2.15 is known as the first-order necessary condition for optimality. A point $\overline{x}^*$ complying this condition could be at optimum, but not necessarily, because it could also be a saddle point. To ensure that $\overline{x}^*$ is at least a local minimum, $\overline{\overline{H}}\,(\overline{x}^*)$ must be positive definite or positive semidefinite. On the other hand, to ensure that $\overline{x}^*$ is at least a local maximum, $\overline{\overline{H}}\,(\overline{x}^*)$ must be negative definite or negative semidefinite.

To solve an unconstrained optimization problem, a gradient-based approach can be used. Such methods basically take an initial solution and start a search for regions where the gradient is reduced. To do that, a search direction and the step size must be determined. The first one indicates in which direction the movement will be performed, and the second one indicates how large the movement will be. The objective is to find a solution for which the gradient is zero, which represents a stationary point, which can be a minimum or a maximum if it complies with the conditions mentioned in the previous paragraph. The general algorithm for a gradient-based method is as follows:

1. Select an initial solution, $\overline{x}_0 = [x_1^0 \; x_2^0 \; \ldots \; x_n^0]^T$, and evaluate the objective function at $\overline{x}_0$.

2. Determine the gradient of the objective **function**, $\nabla f\,(\overline{x}) = \left[\frac{\partial f}{\partial x_1} \; \frac{\partial f}{\partial x_2} \; \cdots \; \frac{\partial f}{\partial x_n}\right]^T$, and, if necessary, the Hessian matrix for the objective function.

3. Establish the convergence criteria. It can be a tolerance value, $\in_1$, for the gradient vector in the current iteration k:

$$\left|\nabla f\left(\bar{x}^k\right)\right| \leq \in_1 \tag{2.16}$$

Other criteria can involve stopping the calculations when the objective function has no important changes from one iteration to another:

$$\left|f\left(\bar{x}^{k+1}\right) - f\left(\bar{x}^k\right)\right| \leq \in_2 \tag{2.17}$$

Finally, a maximum number of iterations can also be fixed.

4. Calculate the search direction for the current iteration, $\bar{S}^k$.
5. Calculate the step size for the current iteration, α^k.
6. Determine the new solution, $\bar{x}^{k+1}$, as follows:

$$\bar{x}^{k+1} = \bar{x}^k + \alpha^k \bar{S}^k \tag{2.18}$$

7. Continue iterating from step 3 to step 5 until the convergence criteria are complied.

There are several gradient-based methods for the solution of unconstrained optimization problems, and the difference among each other is the way to compute the search direction and the step size. In Table 2.2, three classic optimization methods are shown: the steepest descent method, the conjugate gradient method, and the Newton's method. The steepest descent method is based only on the information provided by the first derivative in the current iteration. The conjugate gradient method uses the information provided by the first derivative in the current iteration and also in the previous iteration, together with the information about the convexity/concavity of the function. The Newton's method uses the information about the first derivative and the Hessian matrix in the current iteration. The last two methods, thus, have a faster convergence than the steepest descent method.

2.5 Equality-Constrained Optimization

Most of the process engineering optimization problems are, indeed, constrained. Thus, it is important to understand how to deal with such situations. In this section, a couple of methods for equality-constrained optimization are discussed. A general way to represent an equality-constrained optimization problem can be obtained by simplifying Equation 1.8:

TABLE 2.2

Gradient-Based Optimization Methods

Steepest Descent	**Conjugate Gradient**
$\bar{S}^0 = -\nabla f(\bar{x}^0)$	$\bar{S}^0 = -\nabla f(\bar{x}^0)$
Solve $(\bar{S}^0)^T \nabla f(\bar{x}^0 + \alpha^0 \bar{S}^0)$	$\alpha^0 = -\dfrac{\nabla^T f(\bar{x}^0)\bar{S}^0}{(\bar{S}^0)^T \bar{\bar{H}}(\bar{x}^0)\bar{S}^0}$
$\bar{S}^{k+1} = -\nabla f(\bar{x}^{k+1})$	$\bar{S}^{k+1} = -\nabla f(\bar{x}^{k+1}) + \bar{S}^k \dfrac{\nabla^T f(\bar{x}^{k+1})\nabla f(\bar{x}^{k+1})}{\nabla^T f(\bar{x}^k)\nabla f(\bar{x}^k)}$
Solve $(\bar{S}^{k+1})^T \nabla f(\bar{x}^{k+1} + \alpha^{k+1}\bar{S}^{k+1})$	$\alpha^{k+1} = -\dfrac{\nabla^T f(\bar{x}^{k+1})\bar{S}^{k+1}}{(\bar{S}^{k+1})^T \bar{\bar{H}}(\bar{x}^{k+1})\bar{S}^{k+1}}$

Newton's method

$$\bar{S}^0 = -[\bar{\bar{H}}(\bar{x}^0)]^{-1}\nabla f(\bar{x}^0)$$

$$\alpha^0 = 1$$

$$\bar{S}^{k+1} = -\left[\bar{\bar{H}}\left(\bar{x}^{k+1}\right)\right]^{-1} \nabla f\left(\bar{x}^{k+1}\right)$$

$$\alpha^{k+1} = 1$$

$$\begin{aligned} &\text{optimize } Z = z(\bar{x}) \\ &\text{s.t.} \\ &h(\bar{x}) = 0 \end{aligned} \tag{2.19}$$

Here,the main concern is to obtain an optimal solution for $z(\bar{x})$, which also complies with the set of equality constraints. Two strategies to ensure that are presented here: the method of Lagrange multipliers and the generalized reduced gradient method.

2.5.1 Method of Lagrange Multipliers

In this method, the optimization problem presented in Equation 2.19 is reformulated to obtain an objective function that involves the original objective, $z(\bar{x})$, and the entire set of equality constraints, $h_i(\bar{x}) = 0$, where $i = 1, 2, \ldots, n$. The resultant expression is known as the Lagrangian function and is expressed as follows:

$$\text{optimize } L = z(\bar{x}) + \sum_{i=1}^{m} \lambda_i h_i(\bar{x}) \tag{2.20}$$

where the variables λ_i are known as the Lagrange multipliers. Solutions of the optimization problem expressed by Equation 2.20 can be obtained through the necessary conditions for the Lagrangian function:

$$\frac{\partial L}{\partial \bar{x}} = \nabla z(\bar{x}) + \sum_{i=1}^{m} \lambda_i \nabla h_i(\bar{x}) = 0 \tag{2.21}$$

$$\frac{\partial L}{\partial \lambda_i} = h_i(\bar{x}) = 0 \tag{2.22}$$

From the necessary conditions, a system of equations of **M × M** is obtained. If a solution for the system can be obtained, that solution will represent a stationary point. To ensure the obtained solution is at least a local minimum, $H[L^*]$ should be positive definite or positive semidefinite. On the other hand, if $H[L^*]$ is negative definite or negative semidefinite, the solution is at least a local maximum. If the Hessian matrix is indefinite, the solution is a saddle point.

Example 2.5: The cost of the construction of a distillation column is to be minimized through the method of Lagrange multipliers. This problem has been proposed in the work of Edgar et al. (2001). The objective function is expressed as follows:

$$C = C_p AN + C_s HAN + C_f + C_d + C_b + C_L + C_x \tag{2.23}$$

Moreover, the objective function is constrained by the following expressions:

$$\frac{L}{D} = \left[\frac{1}{1 - (N_{min}/N)}\right]\left(\frac{L}{D}\right)_{min} \tag{2.24}$$

$$A = K(L + D) \tag{2.25}$$

$$C_L = 5000 + 0.7L \tag{2.26}$$

In Equations 2.23 through 2.26, C is the total cost (USD), C_p is the cost per square foot of tray area (USD/ft^2), A is the cross-sectional area of the column (ft^2), N is the number of trays, N_{min} is the minimum number of trays, C_s is the cost of the shell (USD/ft^3), H is the distance between trays (ft), C_f is the cost of the feed pump (USD), C_d is the cost of the distillate pump (USD), C_b is the cost of the bottoms pump (USD), C_L is the cost of the reflux pump (USD), and C_x are the other fixed costs (USD). To solve the problem, Edgar et al. (2001) proposed the values of parameters shown in Table 2.3.

The optimization problem can be formulated in a simplified way by substituting known parameters in the objective function and the constraints, reducing the problem to the following expression:

$$\min C = 50AN + C_L + 17{,}000$$

s.t.

$$L - \frac{1000N}{N-5} = 0 \tag{2.27}$$

$$A - 0.01L - 10 = 0$$

$$C_L - 0.7L - 5000 = 0$$

It can be observed that all the constraints have been modified into the standard form $h_i(\bar{x}) = 0$. Now, we can start the solution of the problem by establishing the optimization problem in terms of the Lagrangian function, Lg.

$$\min \text{Lg} = 50AN + C_L + 17{,}000 + \lambda_1\left(L - \frac{1000N}{N-5}\right) + \lambda_2(A - 0.01L - 10) + \lambda_3(C_L - 0.7L - 5000) \tag{2.28}$$

The nomenclature for the Lagrangian function is slightly modified in this problem to avoid confusion with the variable L. The necessary conditions are as follows:

$$\frac{\partial \text{Lg}}{\partial A} = 50N + \lambda_2 = 0 \tag{2.29}$$

$$\frac{\partial \text{Lg}}{\partial N} = 50A - \lambda_1\left[\frac{-5000}{(N-5)^2}\right] = 0 \tag{2.30}$$

TABLE 2.3

Parameter Data for Example 2.5

Parameter	Value	Parameter	Value	Parameter	Value
C_p	30	C_d	3000	D	1000
C_s	10	C_b	2000	N_{min}	5
H	2	C_x	8000	$(L/D)_{min}$	1
C_f	4000	F	1500	K	0.01

$$\frac{\partial Lg}{\partial C_L} = 1 + \lambda_3 = 0 \tag{2.31}$$

$$\frac{\partial \text{Lg}}{\partial L} = \lambda_1 - 0.01\lambda_2 - 0.7\lambda_3 = 0 \tag{2.32}$$

$$\frac{\partial \text{Lg}}{\partial \lambda_1} = L - \frac{1000N}{N - 5} = 0 \tag{2.33}$$

$$\frac{\partial \text{Lg}}{\partial \lambda_2} = A - 0.01L - 10 = 0 \tag{2.34}$$

$$\frac{\partial \text{Lg}}{\partial \lambda_3} = C_L - 0.7L - 5000 = 0 \tag{2.35}$$

A system with seven equations and seven variables is obtained. Equation 2.30 contains quadratic terms, and therefore multiple solutions can be expected. By solving the system of equations, the following feasible solution is obtained:

$$N^* = 9$$

$$L^* = 2250$$

$$A^* = 32.5$$

$$C_L^* = 6575$$

$$C^* = 38,200$$

The previous solution is a stationary point. To ensure it is a minimum, it is necessary to obtain the characteristic values of the Hessian matrix for the Lagrangian function. Nevertheless, because there are bilinear terms in the cost function (i.e., the product *AN*), it is expected that the objective function is nonconvex. This can be easily proved by computing the characteristic values of the objective function, which are $\lambda_1^{EIG} = -50$, $\lambda_2^{EIG} = 0$, and $\lambda_3^{EIG} = 50$. Thus, the solution obtained for the constrained problem is expected to be a saddle point.

2.5.2 Generalized Reduced Gradient Method

When solving an equality-constrained optimization problem, we are looking for a solution $\bar{x}$ that provided us the minimum (or the maximum) value for the objective function, but strictly complying with the constraints. In some cases, the objective function can be directly modified to include all the constraints. In the generalized reduced gradient method, the constraints are manipulated to put some of the variables of the problem as function of other variables, then replacing those variables in the objective function for those generated functions. Thus, an unconstrained problem, or at least an equality-constrained problem with a reduced number of constraints, can be obtained. The modified objective function could be a single-variable function; it should then be solved by the basic principles of calculus. Otherwise, it can be a multivariable function with no constraints. Thus, a gradient-based method may be useful to solve the reduced problem. In other cases, it is not possible to obtain explicit functionalities for all the variables, and some equality constraints could not be included in the objective function. Nevertheless, the modified problem will have a reduced number of equality constraints, and it could be solved by using a method for equality-constrained optimization problems, such as the method of Lagrange multipliers.

Example 2.6: Solve Example 2.5 using the generalized reduced gradient method.

We can recall that the optimization problem, once the values given in Table 2.3 have been replaced in the objective function, is expressed as follows:

$$\min C = 50AN + C_L + 17{,}000$$

s.t.

$$L - \frac{1000N}{N-5} = 0 \tag{2.36}$$

$$A - 0.01L - 10 = 0$$

$$C_L - 0.7L - 5000 = 0$$

From Equation 2.36, it can be observed that the first constraint can be modified to obtain

$$L = \frac{1000N}{N-5} \tag{2.37}$$

Similarly, the second and third constraints can be modified to obtain

$$A = 0.01L + 10 \tag{2.38}$$

$$C_L = 0.7L + 5000 \tag{2.39}$$

It is clear that L is a function of the number of trays in the column. Moreover, the cross-sectional area A and the cost of the reflux pump C_L are both functions of L and, as a consequence, are also functions of the number of trays. Replacing Equations 2.38 and 2.39 in the objective function, the optimization problem can be stated as follows:

$$\min C = 50\,(0.01L + 10)N + 5000 + 0.7L + 17{,}000$$

$$\text{s.t.} \tag{2.40}$$

$$L = \frac{1000N}{N-5}$$

Equation 2.40 represents an optimization problem with one equality constraint. Nevertheless, such constraint can be used to put all the objective functions in terms of the number of trays, as follows:

$$\min C = 50\left[0.01\left(\frac{1000N}{N-5}\right) + 10\right]N + 5000 + 0.7\left(\frac{1000N}{N-5}\right) + 17{,}000 \tag{2.41}$$

Rearranging Equation 2.41, the problem is reduced to the following expression:

$$\min C = \frac{1000N^2 - 1800N}{N-5} + 22{,}000 \tag{2.42}$$

The optimization problem shown in Equation 2.42 is a single-variable problem; thus, it can be easily solved by using the basic concepts of calculus. The first derivative of the objective function can be obtained as follows:

$$\frac{dC}{dN} = \frac{(N-5)(2000N - 1800) - (1000N^2 - 1800N)}{(N-5)^2} \tag{2.43}$$

If Equation 2.43 is equaled to zero and solved, two solutions are obtained. The first one indicates that $N^* = 1$; nevertheless, this is not a feasible solution because the number of trays cannot be smaller than N_{min}. The second solution indicates that $N^* = 9$, which is a feasible solution. The information about the other variables can be obtained from the functions generated from the original constraints, and the final solution can be given as follows:

$$N^* = 9$$

$$L^* = 2250$$

$$A^* = 32.5$$

$$C_L^* = 6575$$

$$C^* = 38{,}200$$

This solution is the same as that obtained through the method of Lagrange multipliers, and, as has been already proved, it is a saddle point.

2.6 Equality- and Inequality-Constrained Optimization

A more general optimization problem involves a feasible region bounded by equality and inequality constraints. A general way to represent an equality- and inequality-constrained optimization problem is as follows:

$$\begin{aligned} &\text{optimize } Z = z(\bar{x}) \\ &\text{s.t.} \\ &h(\bar{x}) = 0 \\ &g(\bar{x}) \leq 0 \end{aligned} \tag{2.44}$$

For these type of problems, an optimal solution for $z(\bar{x})$ that complies with both the equality and the inequality constraints must be obtained. The equality constraints must always be satisfied as $h(\bar{x}) = 0$. Nevertheless, the inequality constraints can be complied in the form $g(\bar{x}) = 0$, for which it is mentioned that the constraint is active; or it can be satisfied in the form $g(\bar{x}) < 0$, and the constraint is inactive. The solution of the optimization problem will depend on the number of active and inactive constraints.

To find a solution to an equality- and inequality-constrained optimization problem, an approach similar to that for the method of Lagrange multipliers can be implemented. A new objective function, which includes all the equality and inequality constraints, can be stated as follows:

$$\text{optimize } L = z(\bar{x}) + \sum_{i=1}^{m} \lambda_i h_i(\bar{x}) + \sum_{j=1}^{n} \mu_j g_j(\bar{x}) \tag{2.45}$$

This function is known as the augmented Lagrangian function. As before, λ_i are the Lagrange multipliers. The new variables μ_j are known as the Karush–Kuhn–Tucker multipliers because of the contributions of William Karush, Harold W. Kuhn, and Albert W. Tucker to the solution method of such problems. The necessary conditions for the augmented Lagrangian function can be stated as follows:

$$\frac{\partial L}{\partial \bar{x}} = \nabla z(\bar{x}) + \sum_{i=1}^{m} \lambda_i \nabla h_i(\bar{x}) + \sum_{j=1}^{n} \mu_j \nabla g_j(\bar{x}) = 0 \tag{2.46}$$

$$\frac{\partial L}{\partial \lambda_i} = h_i(\bar{x}) = 0 \tag{2.47}$$

$$\frac{\partial L}{\partial \mu_j} = g_j(\bar{x}) = 0 \tag{2.48}$$

The set of necessary conditions is, once more, a system of equations of **M × M**. Nevertheless, if we try to obtain a solution, it may not exist, because such conditions imply that all the inequality constraints are active, which may not be true. Therefore, it is necessary to detect which of the inequality constraints are active. The active set strategy is a useful tool to obtain such information.

2.6.1 Active Set Strategy

As aforementioned, when solving the necessary conditions for the augmented Lagrangian function, it is necessary to know which of the inequality constraints are active, because only those constraints must be included in the function L. To detect the active inequalities, the active set strategy can be used. The steps of the method are explained in this section.

1. Set all the Karush–Kuhn–Tucker multipliers to zero. This implies that all the inequality constraints are inactive.
2. Solve the system of equations given by the necessary conditions. This provides the intermediate solution $\bar{x} = \bar{x}^*_{\text{INT}}$.
3. If for any j, all the inequality constraints are satisfied at the solution found in step 2, i.e., $g_j\left(\bar{x}^*_{\text{INT}}\right) \le 0$, and all the Karush–Kuhn–Tucker multipliers are positive, then an optimal solution has been found, and $\bar{x}^* = \bar{x}^*_{\text{INT}}$.
4. If one or more of the inequality constraints are not satisfied at the solution $\bar{x}^*_{\text{INT}}$, or one or more μ_j are negative, the solution found in step 2 is outside the feasible region. Thus, the constraint with the largest violation,$g_k(\bar{x})$, is turned into active and added to the augmented Lagrangian function.

The steps for the active set strategy are represented as a flowchart in Figure 2.7. An example is presented to show the application of this methodology.

Example 2.7: Solve the problem presented in Example 2.5. Nevertheless, to avoid unfeasible solutions where the number of stages is equal or less than $N_{\min}$, the following inequality constraint should be added:

$$N \ge N_{\min} + 1 \tag{2.49}$$

To solve this problem, the inequality constraint is first modified into the standard form $g(\bar{x}) \le 0$ as follows:

$$6 - N \le 0 \tag{2.50}$$

where the numerical value of $N_{\min}$ has been already substituted. The augmented Lagrangian function is then stated as follows:

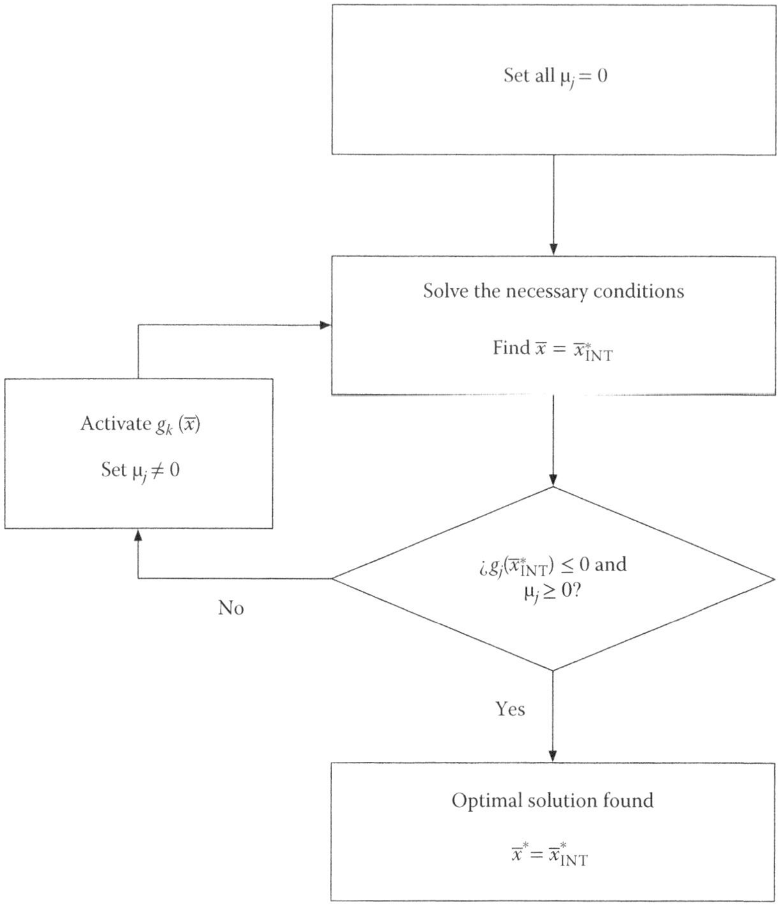

FIGURE 2.7
Active set strategy.

$$\min \mathrm{Lg} = 50AN + C_L + 17{,}000 + \lambda_1\left(L - \frac{1000N}{N-5}\right) + \lambda_2\left(A - 0.01L - 10\right) + \lambda_3\,(C_L - 0.7L - 5000) + \mu_1\,(6 - N) \tag{2.51}$$

To start with the active set strategy, let us suppose the Karush–kuhn–Tucker multiplier is zero, i.e., $\mu_1 = 0$. Then, the necessary conditions are the same as those represented by Equations 2.29 through 2.35, and the solution is given by $N^* = 1$ and $N^* = 9$. Both the solutions are analyzed.

The inequality constraint evaluated in the intermediate solution $N^* = 9$ gives $6 - 9 = -3$ and $-3 < 0$. Thus, for $N^* = 9$, the inequality constraint is complied, and the solution is a stationary point for the problem constrained by equalities and inequalities. The entire solution is expressed as follows:

$$N^* = 9$$

$$L^* = 2250$$

$$A^* = 32.5$$

$$C_L^* = 6575$$

$$C^* = 38,200$$

If the intermediate solution $N^* = 1$ is substituted in the inequality constraint, it results in $6 - 1 = 5$, and $5 > 0$. This implies that $N^* = 1$ does not satisfy the inequality constraint, and it must be activated. Then, Equation 2.51 is used for generating the new set of necessary conditions:

$$\frac{\partial \mathrm{Lg}}{\partial A} = 50N + \lambda_2 = 0 \tag{2.52}$$

$$\frac{\partial \mathrm{Lg}}{\partial N} = 50A - \lambda_1 \left[\frac{-5000}{(N-5)^2} \right] - \mu_1 = 0 \tag{2.53}$$

$$\frac{\partial \mathrm{Lg}}{\partial C_L} = 1 + \lambda_3 = 0 \tag{2.54}$$

$$\frac{\partial \mathrm{Lg}}{\partial L} = \lambda_1 - 0.01\lambda_2 - 0.7\lambda_3 = 0 \tag{2.55}$$

$$\frac{\partial \mathrm{Lg}}{\partial \lambda_1} = L - \frac{1000N}{N-5} = 0 \tag{2.56}$$

$$\frac{\partial \mathrm{Lg}}{\partial \lambda_2} = A - 0.01L - 10 = 0 \tag{2.57}$$

$$\frac{\partial \mathrm{Lg}}{\partial \lambda_3} = C_L - 0.7L - 5000 = 0 \tag{2.58}$$

$$\frac{\partial \mathrm{Lg}}{\partial \mu_1} = 6 - N = 0 \tag{2.59}$$

By solving Equations 2.52 through 2.59, the following solutions are obtained:

$$N^* = 6$$

$$L^* = 6000$$

$$A^* = 70$$

$$C_L^* = 9200$$

$$C^* = 47,200$$

This solution complies with the inequality constraint; thus, it is a second stationary point. Nevertheless, it is clear that the first solution, where $N^* = 9$, provides a better value of the objective function.

2.7 Software for Deterministic Optimization

In this chapter, some of the basic concepts of deterministic optimization have been presented. In particular, methods for the solution of nonlinear optimization problems are presented. Nevertheless, deterministic optimization embraces several other types of problems, such as the mixed-integer optimization problems or the general disjunctive optimization problems. Furthermore, process engineering models usually involve a great number of constraints, which make finding solutions for the models difficult. Because of that, the use of software for the solution of such models, and the associated optimization problems, is mandatory. Deterministic optimization software, such as GAMS and LINDO, are available in the market. These software use an equation-based approach. They are based on the use of solvers to determine optimal solutions for the objective function subject to a set of constraints. Solvers are basically routines to optimize, and most of them are based on gradient methods. Both GAMS and LINDO have the capacity to deal with different optimization problems, such as linear programming (LP), nonlinear programming (NLP), mixed-integer linear programming (MILP), and mixed-integer nonlinear programming (MINLP), using local or global solvers. The user is required to write the model to be solved, and the software uses a given method to optimize it in terms of the objective function. In fact, although the software uses default solvers for each type of optimization problem, the user must be careful to properly select the solver.

Deterministic optimization software can be used to solve process engineering problems when the model is available. Additional strategies can be required to make easier finding an optimum for nonconvex solution spaces and/or avoiding falling into local optimum. Nevertheless, those strategies, along with the guidelines for the use of deterministic optimization software, are beyond the scope of this book. For a deeper knowledge of GAMS, the reader is referred to the user's manuals (Brooke et al., 1998; McCarl, 2004).

References

L.T. Biegler, 2010, *Nonlinear Programming: Concepts, Algorithms, and Applications to Chemical Processes*, Philadelphia, PA: MOS-SIAM.

A. Brooke, D. Kendrick, A. Meeraus, R. Raman, R.E. Rosenthal, 1998, *GAMS: A User's Guide*, Washington, DC: GAMS Development Corporation.

U. Diwekar, 2010, *Introduction to Applied Optimization*, 2nd edition, New York, NY: Springer.

T.F. Edgar, D.M. Himmelblau, L.S. Lasdon, 2001, *Optimization of Chemical Processes*, 2nd edition, New York, NY: McGraw-Hill.

D. Liberzon, 2012, *Calculus of Variations and Optimal Control Theory: A Concise Introduction*, Princeton, NJ: Princeton University Press.

B.A. McCarl, 2004, *GAMS User's Guide: 2004*, Washington, DC: GAMS Development Corporation, available at www.gams.com. Accessed on 29 March 2016.

D.A. Pierre, 1986, *Optimization Theory with Applications*, New York, NY: Dover Publications.

A. Shapiro, D. Dentcheva, A. Ruszczynski, 2014, *Lectures on Stochastic Programming: Modeling and Theory*, 2nd edition, Philadelphia, PA: MOS-SIAM.

3

Stochastic Optimization

3.1 Introduction to Stochastic Optimization

The term "stochastic" can be related with randomness or uncertainty. Stochastic optimization is, indeed, based on the concept of random search. Finding an optimum for a given function, thus, implies testing different solutions for the decision variables and evaluating the objective function for each of those test points until a good solution is found. That "good solution" will be expected to be the minimum (or the maximum) of the function, or at least to be close enough to it. Nevertheless, to ensure a solution close to the optimum, a random search is not sufficient. It is necessary to establish criteria to allow avoiding bad solutions as the algorithm advances, so that the final solution is, at least, close to the global optimum. This is known as a directed random search.

It is important to clearly distinguish between stochastic optimization (also known as meta-heuristic optimization) and stochastic programming. Stochastic optimization is a set of methods to solve optimization problems, which are based on a directed random search. However, stochastic programming is a subset of optimization problems in which algebraic constraints with uncertainties are involved. Methods have been reported to solve stochastic programming problems using deterministic strategies (Diwekar, 2010; Shapiro et al., 2014), but they are not the focus of this book. In this book, stochastic (meta-heuristic) optimization methods are used to solve optimization problems in process engineering.

3.2 Stochastic Optimization vs. Deterministic Optimization

Stochastic and deterministic methods are usually considered as opposite approaches for optimization because of the differences on the basis from which the method is developed. First, the deterministic methods are based on rigorous mathematical principles, mainly on the concepts of calculus. As seen in Chapter 2, most of the methods rely on obtaining solutions for which the gradient is zero, and the evaluation of a given solution to determine if it is indeed an optimum is based on the calculation of the second derivatives, in the form of the Hessian matrix. However, stochastic optimization is based on the evaluation of the objective function in the entire feasible region and the comparison of different solutions to select the best solution for each iteration. Nevertheless, occasionally some bad solutions can be selected in a given iteration, depending on some selection probabilities. For convex functions, deterministic methods always ensure finding a global optimum because they are formulated to search for solutions that comply with the optimality conditions. A stochastic optimization method may reach the global optimum or, at least, solutions close to it, even for highly nonconvex functions. This will depend on proper tuning of the parameters of the algorithm. Another difference between both the methods relies on the importance of initial solutions. Local deterministic methods have a strong dependence on initial values, because the selection of that point at the initial iteration may take the solution to a local optimum, depending, once more, on the convexity (or nonconvexity) of the function. On the contrary, the dependence on the initial solution for a stochastic method is not that strong. The main issue is that, if the initial values are not good, the method will require a higher number of iterations to reach a region close to the global optimum. Finally, the computational time and capacity required for the solution of an optimization problem through deterministic methods are relatively low, whereas these are higher for a stochastic method because a wide range of potential solutions are evaluated. Nonetheless, stochastic methods are a good alternative when dealing with highly nonconvex problems with a high number of degrees of freedom, reducing the difficulties on finding feasible initial solutions, and avoiding the necessity of computing the derivatives, which can be a difficult task for complex functions. Moreover, the stochastic methods can deal with problems for unknown models considering only the input–output data, following a gray-box approach.

3.3 Stochastic Optimization with Constraints

The aforementioned generalities about stochastic optimization are valid for the solution of unconstrained problems. Nevertheless, most of the engineering problems have a set of constraints associated with the objective functions, given by the model of the studied system, and must be considered to obtain feasible solutions. The basic stochastic optimization methods cannot deal with constrained problems; thus, strategies have been developed to allow solving such problems. The strategies used to handle constraints in stochastic optimizations can be classified as follows: penalty functions, special representations and operators, repair algorithms, separation of objectives and constraints, and hybrid methods (Coello Coello, 2002). One of the most used constraints-handling methods comprises the use of penalty functions. This strategy is explained in this section.

One of the first reports on the use of penalty functions to deal with constrained optimization problems was presented by Carroll (1961). In general, the method involves adding or subtracting a certain quantity to the objective function, depending on how big is the violation to the constraints. Thus, the constrained problem is converted into an unconstrained one. Because most of the meta-heuristic optimization methods involve the selection of the best solution for each iteration, if the objective function of a given solution is worsened because it violates one or more constraints, the probability of selecting such solution as a good one is reduced, and the solutions satisfying all the constraints are more likely to be chosen.

One of the basic approaches to implement the penalty function involves the use of exterior penalties. Such methods can start out of the feasible region, and then move into it. This is one of their main advantages because no feasible initial solution is required. According to Coello Coello (2002), the formulation for an exterior penalty function is given as follows:

$$\phi(\bar{x}) = \mathrm{f}(\bar{x}) \pm \left[\sum_{i=1}^{n} r_i \cdot G_i + \sum_{j=1}^{p} c_j \cdot L_j \right] \tag{3.1}$$

where $\mathrm{f}(\bar{x})$ is the original objective function and $\phi(\bar{x})$ is the expanded objective function; G_i is the function of the inequality constraints, $g_i(\bar{x})$, whereas L_j is the function of the equality constraints, $h_i(\bar{x})$; and finally, r_i and c_j are known as penalty factors and are positive constants. The penalty functions G_i and L_j must be selected to avoid too low or too high penalizations to the objective function $\mathrm{f}(\bar{x})$. In general, the penalty functions can be stated as follows (Yeniay, 2005):

$$G_i = \max\left[0, g_i(\bar{x})\right]^{\beta} \tag{3.2}$$

$$L_j = \left|h_j(\bar{x})\right|^{\gamma} \tag{3.3}$$

where β and γ are usually set as 1 or 2. The penalty factors can be calculated in several ways: keeping them constant in all the optimization procedure (static penalties), computing them in terms of the number of iterations (dynamic penalties), using annealing approaches, among others. For a detailed description of the particular penalty methodologies, the reader is referred to the work of Coello Coello (2002).

3.4 Genetic Algorithms

Genetic algorithms are among the first developed stochastic optimization methods. They were proposed by Holland (1975) and are a subtype of the so-called evolutionary algorithms. They emulate the evolution of a species, where the more capable individuals in a given population have higher chances to pass their genes to further generations. In mathematical terms, a solution $\bar{x}$ is an individual in the GA, and a given set of solutions forms a population. One of the particularities of the GAs is that the solutions are represented in terms of chromosomes, i.e., chains containing the genetic information of each individual. That codified representation is known as genotype, whereas the "manifestation" of the genotype, i.e., the physical/mathematical system, is known as phenotype. Some examples of chromosomes are shown in Figure 3.1, where binary, integer, and alphabetic representations can be observed. The type of representation to be used depends on the problem to be solved. Each locus (position on the chain) can show different values (alleles), and each combination of locus and values represents different solutions for the objective function $f(\bar{x})$. For engineering applications, codification with real numbers is more advisable because of the similarity between the genotype and the phenotype spaces (Gen and Cheng, 2000).

Once the problem has been codified, an initial solution is required to start the algorithm. The GAs function with a set of solutions in a simultaneous way, thus the initial point is indeed a population of solutions, each one with particular characteristics (i.e., different genetic information) that differentiate it from the others. The initial population is generated randomly. Then, it is necessary to evaluate the individuals on that first generation of solutions to determine which of them are good individuals and which are bad individuals. This classification is given by the so-called fitness function, which is strongly related to the objective function. Thus,

1	0	1	1	0	0	1	0	0	1	1

0	5	9	0	3	2	8	7	1	5	4

A	R	D	T	Y	L	M	F	I	G	C

FIGURE 3.1
Codification in a genetic algorithm.

for minimization, "good individuals" are those with a low value of $f(\bar{x})$. Then, a selection procedure is started. In this step, some of the individuals in the generation are selected to reproduce and give birth to the next generation, which is expected to have better characteristics than those of the previous generation. In general, the best individuals of the generation (i.e., those with the better values of $f(\bar{x})$ have higher probabilities of being selected for reproduction). Nevertheless, other individuals can also be selected to give genetic variability in the following generation, ensuring that a wider space in the feasible region is analyzed.

The genetic operator of crossover is applied to all the selected individuals. By this operator, two random individuals are selected and their genetic information is combined to give birth to two new individuals, which have part of the genetic information of each parent, but the phenotype corresponds to a different solution. A second genetic operator, known as mutation, can be applied to a small proportion of the offspring (i.e., a number of individuals in the new generation). Through mutation, one or more of the alleles are modified in a given locus to produce a new solution, different from the one obtained through crossover. This operator is helpful to amplify the search space on the algorithm. Nevertheless, mutation probability must be kept small to avoid losing good solutions. A simplified representation of crossover and mutation is presented in Figure 3.2 using a binary codification. In Figure 3.2a, the crossover is applied to two parents, and two new individuals (the offspring) are obtained. In Figure 3.2b, the genetic information of one individual is randomly modified, giving birth to a new, mutated individual. Once the new generation has been produced, it is evaluated, and the procedure starts again, until the convergence criterion (CC) is accomplished. This criterion may be defined by fixing a maximum number of generations. The number of generations must be tuned to ensure that the algorithm converges to a region close to the global optimum. Other alternative for CC is to stop the iterations when the objective function for all the individuals in a given generation shows no representative differences. Figure 3.3 shows a graphical representation of a GA.

3.5 Differential Evolution

Differential evolution is an evolutionary method. It shares some characteristics with the GAs. It was first proposed by Storn and Price (1997) as a strategy to solve the Chebyshev polynomial fitting problem (Shaoqiang et al., 2010). Similar to the GAs, the DE method functions with generations, where each generation comprises a number of parameter vectors representing a set of solutions. It is important to recall that DE uses a real-number representation at the parameter vectors. To start with the algorithm, an initial solution is

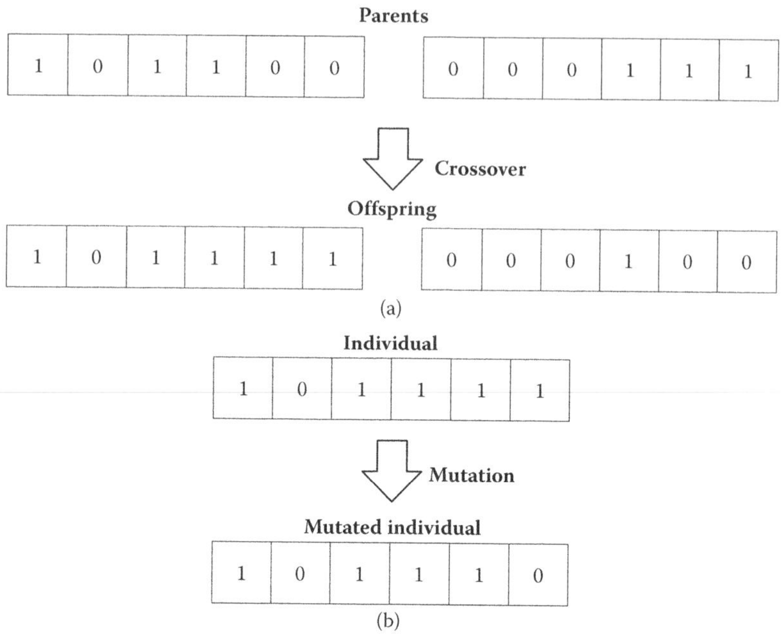

FIGURE 3.2
Genetic operators: (a) crossover and (b) mutation.

randomly generated. The solutions in the generation are evaluated through the fitness function, which is related to the objective function. To produce the next generation, an individual $\overline{x}_{SEL}$ is randomly selected as candidate to be substituted through the crossover operation. Here, three other individuals $(\overline{x}_{p1}, \overline{x}_{p2}, \overline{x}_{p3})$ are selected as parents, where one of the individuals will act as the main parent. Then, a fraction of the difference between the values of each variable in the other two parents is computed. Those values are added to the value of the respective variable in the main parent, which can be expressed as follows:

$$(\overline{x}_{NEW})^T = (\overline{x}_{p1})^T + F \times [(\overline{x}_{p2})^T - (\overline{x}_{p3})^T] \tag{3.4}$$

where F is a randomly generated number and $F \in (0,1)$ (Abbass et al., 2001). The new individual $\overline{x}_{NEW}$ is then compared with the selected individual. If $f(\overline{x}_{NEW})$ is better than $f(\overline{x}_{SEL})$, then $\overline{x}_{SEL}$ is replaced by $\overline{x}_{NEW}$ in the population. Otherwise, $\overline{x}_{SEL}$ remains as an individual for the next generation. This operation takes place until the new generation has been completed. The procedure continues until the CC has been reached, which may imply a maximum number of generations. Figure 3.4 shows a graphical representation of the DE method.

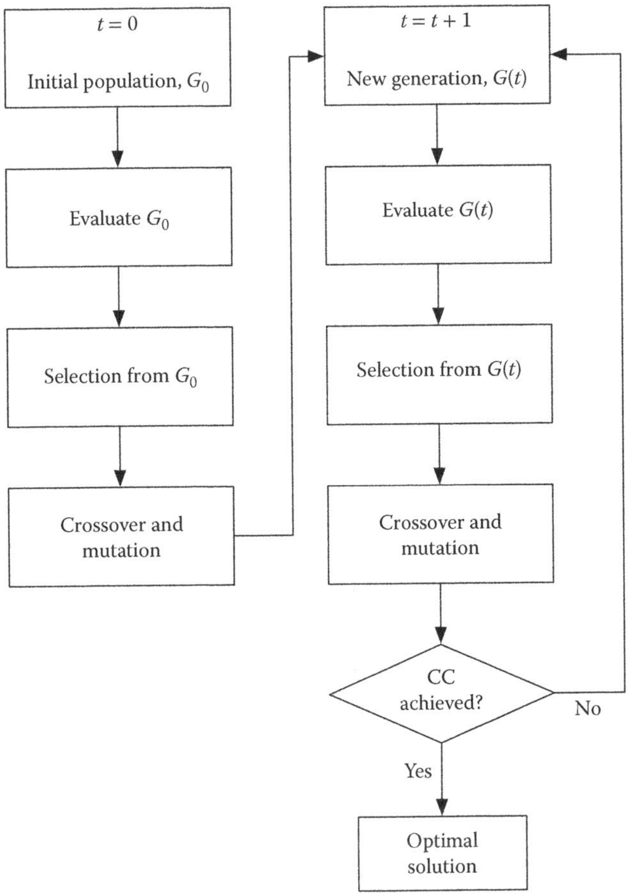

FIGURE 3.3
Block diagram for a genetic algorithm.

3.6 Tabu Search

Tabu search is an optimization method proposed by Glover (1977, 1989). The method was originally developed to solve combinatorial problems related with scheduling and covering (Glover, 1989). One of the most important concepts, from which the method takes its name, is the so-called tabu list. The tabu list consists of a set containing some of the solutions that have been already proved. In general, the method starts with a single initial solution $\overline{x}$, setting the tabu list as empty. The initial solution is then perturbed several times to generate a number of new solutions $\overline{x}'$, which is known as the neighborhood of $\overline{x}$, $N(\overline{x})$. This neighborhood can be obtained by applying

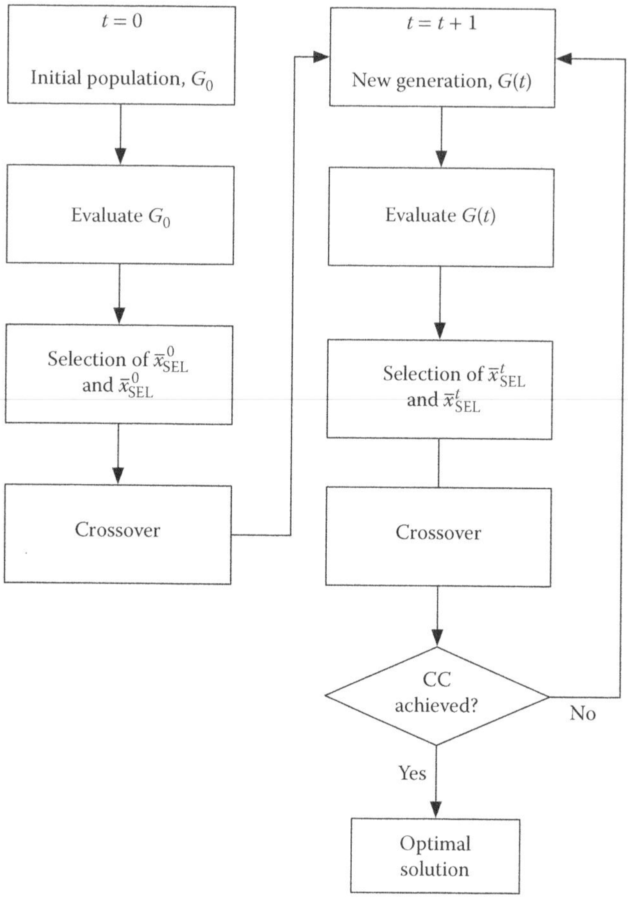

FIGURE 3.4
Block diagram for the differential evolution method.

a modification m to the initial solution, i.e., $\bar{x}' = \bar{x}' \pm m$ (Fiechter, 1994). At the first steps of the algorithm, it is possible to move to a solution $f(\bar{x}')$, no matter if it is better than $f(\bar{x})$ or not. A given number of the last obtained solutions is then added to the tabu list. The solutions in the previous iteration are then compared and the best one is selected as the new suboptimal. The new solution is then perturbed to generate another set of alternative solutions. For a next iteration, if a new solution is contained in the tabu list, it must be rejected and an alternative solution is proposed. As the iterations advance, the oldest components of the tabu list are deleted. The method continues until the stop criterion is reached, which may imply a maximum number of iterations. Figure 3.5 presents a graphical representation of the tabu search method.

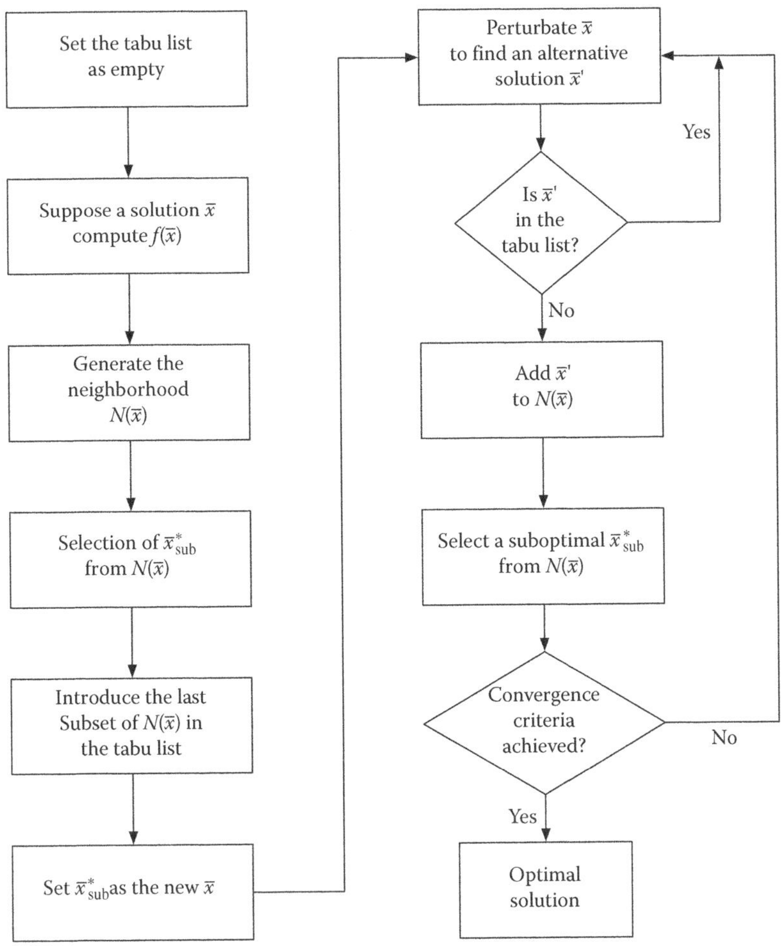

FIGURE 3.5
Block diagram for the tabu search method.

3.7 Simulated Annealing

Kirkpatrick et al. (1983) established that there is a similarity between the behavior of a system reaching thermal equilibrium and the performance of an optimization procedure. This was considered the basis for the creation of the simulated annealing approach, which emulates the phenomena of annealing in solids. In the annealing procedure, a solid is first heated at high temperature. At this stage, the energy of the system is quite high, and the atoms in the solid are randomly distributed. Then, the temperature is slowly reduced until a new equilibrium is reached. This procedure continues until the atoms are ordered in a crystalline structure, where the energy of the system is at its minimum. The solid treated with annealing is quite resistant. However, if the temperature was not gradually reduced, a thermal shock is induced, and the solid becomes fragile. In the simulated annealing method, a simulated temperature is used to control the algorithm. As in the physical phenomena, this temperature must be high to cause a random behavior. Once the initial simulated temperature is selected, an initial solution $\bar{x}$ is proposed, and the value of the objective function $f(\bar{x})$ is computed. Then, a second solution $\bar{x}'$ is proposed, computing the value of $f(\bar{x}')$. The two solutions are compared and, if $f(\bar{x}')$ is better than $f(\bar{x})$ (i.e., if, for a minimization, $f(\bar{x}') \geq f(\bar{x})$), then $\bar{x}'$ is selected as the new solution. If $f(\bar{x}')$ is worse than $f(\bar{x})$, it is not immediately discarded. Instead, a probability of selection is computed using the Metropolis formula (Metropolis et al., 1953):

$$P(\Delta) = \exp\left(-\frac{\Delta}{T}\right) \tag{3.5}$$

where

$$\Delta = f(\bar{x}') - f(\bar{x}) \tag{3.6}$$

If the probability of selection is higher than a random number a, then $\bar{x}'$ is selected as the new solution. Otherwise, the method returns to the previous proposal, $\bar{x}$. This implies that, if a given solution is "bad," it still has probabilities of being selected as a new solution. This is helpful to perform a search on all feasible regions, avoiding local optimum. The proposal and selection of new solutions continue until a stationary point is reached. Then, the temperature is decreased, and a new set of proposals is established. As the temperature decreases, the probability of selection is lower. Thus, when the algorithm advances, the "bad" solutions have less chances to be selected because the method is expected to be converging to the global optimum. The algorithm stops when the freezing temperature (T_{freeze}) is reached, where the solution is stable and the same solution is selected among a certain number of proposals. Figure 3.6 shows a graphical representation of the simulated annealing.

3.8 Other Methods

The described methods are examples of stochastic optimization strategies, but there are several other alternatives to solve optimization problems using meta-heuristic approaches. The GAs and DEs are methods based on the analysis of populations (or sets of solutions), where tabu search and simulated annealing are methods based on the analysis of single solutions. In this

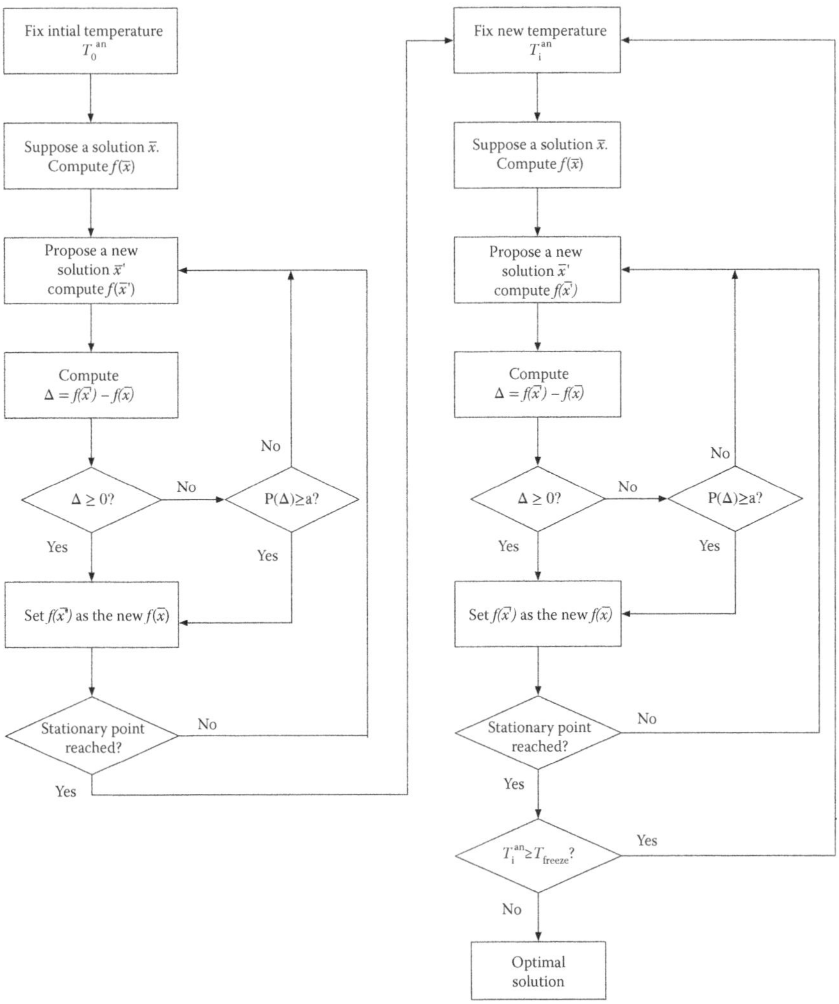

FIGURE 3.6
Block diagram for the simulated annealing method.

section, other stochastic optimization methods are mentioned and a brief description of each method is presented.

3.8.1 Ant Colony Optimization

The ant colony optimization method was developed by Dorigo and Gambardella (1997) and is inspired by the behavior of ant colonies. In general, the ants indicate the way from their nest to a food source by depositing pheromones on the ground. When various ants are taking different paths, the path representing the shortest distance will have a higher concentration of pheromones, thus most of the ants will be attracted to follow that route and will increase the concentration of pheromones even more. In the end, all the ants nearby will follow the shortest path. The ant colony optimization method emulates this behavior. Two parameters are of importance to the algorithm: the pheromone value and the age of a solution. The algorithm starts with a randomly generated set of solutions. Then, a local search procedure starts, where a simulated ant selects a solution, in terms of the pheromone values. Then, in terms of the age of the current solution, a new solution is selected. If the fitness function of the new solution is better than that of the previous solution, the new solution is selected. Otherwise, the previous solution remains. In both cases, the values of age and pheromone are modified. Then, a global search takes place and the pheromone values are updated to consider the phenomena of pheromone evaporation. The procedure continues until the CC is achieved (Jayaraman et al., 2010). This method has been used for applications such as the optimization of project scheduling (Merkle et al., 2002), the optimization of water distribution systems (Maier et al., 2003), and the scheduling of batch processes (Jayaraman et al., 2010). Nevertheless, to the best of the authors' knowledge, there are only a few applications of this algorithm to the optimization of chemical processes.

3.8.2 Particle Swarm Optimization

This optimization method is based on the social behavior of animal species, particularly that of human beings. The method was proposed by Kennedy and Eberhart (1995), and it is related to both artificial life and evolutionary programming. It has been developed from the observations and simulations of Reynolds (1987) not only about the movement patterns of flying animals (birds), land animals, and water animals (fishes), but also considering the abstractness that characterizes the human decision-making. In general, the algorithm functions with a set of entities known as particles. An initial population of particles is first generated, and a position and a velocity are assigned to each particle, where the position is given by a solution for the optimization problem. Then, the fitness function is evaluated for each position, and the best value is selected. Then, the position and the velocity of each particle are updated, which implies a movement of the particle in the direction of the best previous solution. The movements continue until a stop criterion is reached (Jarboui et al., 2010). This approach has been applied to the dynamic analysis of reactive systems (Ourique et al., 2002), the parameter estimation of a polypropylene reactor (Martinez Prata et al., 2009), and the optimal design of heat exchangers (Patel and Rao, 2010).

3.8.3 Harmony Search

Harmony search is a method that emulates the musical harmony and was first proposed by Geem et al. (2001). In this approach, the values of the variables are compared with the pitches of instruments, and the objective function is related to an aesthetic standard given by the combined sounds of different instruments. To enhance the objective functions, iterations are required to find better values of the variables, whereas enhancing the aesthetic standard implies practice to obtain best combination of sounds. Finally, the global optimum is represented through the best performance of the musical piece, which is known as a fantastic harmony. In general, the algorithm starts with the specification of the required parameters, which are basically the harmony memory size, the maximum number of improvisations, and the harmony memory considering rate. Then, the harmony memory, which stores solution vectors, is randomly initialized, and the objective function is computed for each solution. Then, the solutions are ordered from the best to the worst. In the next step, a new solution is obtained by improvising the harmony, where one of three operations may occur: harmony memory considering (selecting a solution previously stored in the harmony memory), pitch adjusting (selecting a solution in the harmony memory, then applying a small modification to that solution), or random playing (setting a solution to any value in the feasible region). The harmony memory is then updated and the procedure is repeated until a stopping criterion is reached (Ingram and Zhang, 2010). The harmony search algorithm has been used for some applications related to process engineering, such as water distribution networks (Geem, 2006), design of pressurized vessels (Mahdavi et al., 2007), design of shell-and-tube heat exchangers (Fesanghary et al., 2009), synthesis of heat exchanger networks (Mohammadhasani Khorasany and Fesanghary, 2009), and design of plate–fin heat exchangers (Yousefi et al., 2013).

References

H.A. Abbass, R. Sarker, C. Newton, 2001, PDE: A Pareto frontier differential evolution approach for multi-objective optimization problems, in *IEEE Conference on Evolutionary Computation*, 27–30 May, Seoul, Korea, 2, 971–978.

C.W. Carroll, 1961, The created response surface technique for optimizing nonlinear restrained systems, *Oper. Res.*, 9, 169–184.

C.A. Coello Coello, 2002, Theoretical and numerical constraint-handling techniques used with evolutionary algorithms: A survey of the state of the art, *Comput. Method. Appl. M.*, 191(11–12), 1245–1287.

U. Diwekar, 2010, *Introduction to Applied Optimization*, 2nd edition, New York, NY: Springer.

M. Dorigo, L.M. Gambardella, 1997, Ant colony system: A cooperative learning approach to the traveling salesman problem, *IEEE Trans. Evolut. Comput.*, 1(1), 53–66.

M. Fesanghary, E. Damangir, I. Soleimani, 2009, Design optimization of shell and tube heat exchangers using global sensitivity analysis and harmony search algorithm, *Appl. Therm. Energy*, 29(5–6), 1026–1031.

C.-N. Fiechter, 1994, A parallel tabu search algorithm for large traveling salesman problems, *Discrete Appl. Math.*, 51(3), 243–267.

Z.W. Geem, 2006, Optimal cost design of water distribution networks using harmony search, *Eng. Optimiz.*, 38(3), 259–280.

Z.W. Geem, J.H. Kim, G.V. Loganathan, 2001, A new heuristic optimization algorithm: Harmony search, *Simulation*, 76(2), 60–68.

M. Gen, R. Cheng, 2000, *Genetic Algorithms and Engineering Optimization*, New York, NY: John Wiley & Sons.

F. Glover, 1977, Heuristics for integer programming using surrogate constraints, *Decis. Sci.*, 8(1), 156–166.

F. Glover, 1989, Tabu search—Part I, *ORSA J. Comput.*, 1(3), 190–206.

J.H. Holland, 1975, *Adaptation in Natural and Artificial Systems*, Cambridge, MA: MIT Press.

G. Ingram, T. Zhang, 2010, An introduction to the harmony search algorithm, in *Stochastic Global Optimization: Techniques and Applications in Chemical Engineering*, G.P. Rangaiah (Ed.), Singapore: World Scientific, pp. 301–335.

B. Jarboui, H. Derbel, M. Eddaly, 2010, Particle swarm optimization for solving NLP and MINLP in Chemical Engineering, in *Stochastic Global Optimization: Techniques and Applications in Chemical Engineering*, G.P. Rangaiah (Ed.), Singapore: World Scientific, pp. 271–300.

V.K. Jayaraman, P.S. Shelokar, P. Shingade, V. Pote, R. Baskar, B.D. Kulkarni, 2010, Ant colony optimization: Details of algorithms suitable for process engineering, in *Stochastic Global Optimization: Techniques and Applications in Chemical Engineering*, G.P. Rangaiah (Ed.), Singapore: World Scientific, 237–269.

J. Kennedy, R.C. Eberhart, 1995, Particle swarm optimization, in *Proceedings of IEEE International Joint Conference on Neural Networks*, Perth, Australia, 4, 1942–1948.

S. Kirkpatrick, C.D. Gelatt Jr., M.P. Vecchi, 1983, Optimization by simulated annealing, *Science*, 220(4598), 671–680.

M. Mahdavi, M. Fesanghary, E. Damangir, 2007, An improved harmony search algorithm for solving optimization problems, *Appl. Math. Comput.*, 188(2), 1567–1579.

H.R. Maier, A.R. Simpson, A.C. Zecchin, et al., 2003, Ant colony optimization for design of water distribution systems, *J. Water Resour. Plan. Manage.*, 129(3), 200–209.

D. Martinez Prata, M. Schwaab, E.L. Lima, J.C. Pinto, 2009, Nonlinear dynamic data reconciliation and parameter estimation through particle swarm optimization: Application for an industrial polypropylene reactor, *Chem. Eng. Sci.*, 64(18), 3953–3967.

D. Merkle, M. Middendorf, H. Schmeck, 2002, Ant colony optimization for resource-constrained project scheduling, *IEEE Trans. Evolut. Comput.*, 6(4), 333–346.

N. Metropolis, A. Rosenbluth, M. Rosenbluth, A. Teller, E. Teller, 1953, Equation of state calculations by fast computing machines, *J. Chem. Phys.*, 21(6), 1087–1092.

R. Mohammadhasani Khorasany, M. Fesanghary, 2009, A novel approach for synthesis of cost-optimal heat exchanger networks, *Comput. Chem. Eng.*, 33(8), 1363–1370.

C.O. Ourique, E.V. Biscaia Jr., J.C. Pinto, 2002, The use of particle swarm optimization for dynamical analysis in chemical processes, *Comput. Chem. Eng.*, 26(12), 1783–1793.

V.K. Patel, R.V. Rao, 2010, Design optimization of shell-and-tube heat exchanger using particle swarm optimization technique, *Appl. Therm. Eng.*, 30(11–12), 1417–1425.

C.W. Reynolds, 1987, Flocks, herds and schools: A distributed behavioral model, *Comput. Graph.*, 21(4), 25–34.

C. Shaoqiang, G.P. Rangaiah, M. Srinivas, 2010, Differential evolution: Method, developments and chemical engineering applications, in *Stochastic Global Optimization: Techniques and Applications in Chemical Engineering*, G.P. Rangaiah (Ed.), Singapore: World Scientific, pp. 203–236.

A. Shapiro, D. Dentcheva, A. Ruszczynski, 2014, *Lectures on Stochastic Programming: Modeling and Theory*, 2nd edition, Philadelphia, PA: MOS-SIAM.

R. Storn, K. Price, 1997, Differential evolution—A simple and efficient heuristic for global optimization over continuous spaces, *J. Global. Optim.*, 11(4), 341–359.

O. Yeniay, 2005, Penalty function methods for constrained optimization with genetic algorithms, *Math. Comput. Appl.*, 10(1), 45–56.

M. Yousefi, R. Enayatifar, A.N. Darus, A.H. Abdullah, 2013, Optimization of plate–fin heat exchangers by an improved harmony search algorithm, *Appl. Therm. Energy*, 50(1), 877–885.

4

The Simulator Aspen Plus®

4.1 Importance of Software for Process Analysis

In process engineering, the simulation, design, and optimization of a chemical process plant, which comprises several processing units interconnected by process streams, are the core activities. These tasks require material and energy balancing, equipment sizing, and costing calculation. A computer package that can accomplish these duties is known as a computer-aided process design package or simply a process simulator (also known as process flowsheeting package, flowsheet simulator, or flowsheeting software). The capabilities of a process simulator include an accurate description of physical properties of pure components and complex mixtures, rigorous models for unit operations, as well as numerical techniques for solving large systems of algebraic and differential equations. By a process simulator, it is possible to obtain a comprehensive computer image of a running process, which is a valuable tool in understanding the operation of a complex chemical plant and on this basis can serve for continuously improving the process or for developing new processes.

The purpose of simulation is to model and predict the performance of a process. It involves decomposition of the process into its constituent units for individual study of performance. The process characteristics (e.g., flow rates, compositions, temperatures, pressures, properties, and equipment sizes) are predicted using analysis techniques, which include mathematical models, empirical correlations, and computer-aided process simulation tools (e.g., Aspen Plus). In addition, process analysis may involve the use of experimental methods to predict and validate performance. Therefore, in process simulation, the process inputs and the flowsheet are given, and we are required to predict process outputs (Figure 4.1). This book focuses on Aspen Plus. It is a computer-aided software that uses the underlying physical relationships (e.g., material and energy balances, thermodynamic equilibrium, and rate equations) to predict process performance (e.g., stream properties, operating conditions, and equipment sizes).

There are several advantages of computer-aided simulation:

1. It allows the designer to quickly test the performance of synthesized process flowsheets and provide feedback to the process synthesis activities.

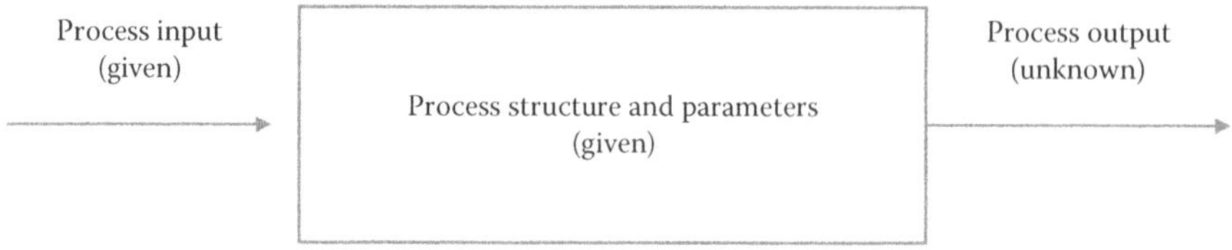

FIGURE 4.1
Process simulation problems.

2. It can be coordinated with process synthesis to develop optimum integrated designs.
3. It minimizes experimental and scale-up efforts.
4. It explores process flexibility and sensitivity by answering "what-if" questions.
5. It quantitatively models the process and sheds insights on process performance.

Following are the important issues to remember before venturing into the exciting world of computer-aided simulation:

1. Do not implicitly trust the results of any simulation tool.
2. Calculated results are only as good as the input you give the simulator.
3. Always convince yourself that the obtained results make physical sense, otherwise you will never be able to convince someone else of the merits of your work.

4.2 Characteristics of the Process Simulator Aspen Plus

The process simulation market underwent severe transformations in the 1985–1995 decade. Relatively few systems have survived; they are CHEMCAD, Aspen Plus, Aspen HYSYS®, PRO/II, ProSimPlus, SuperPro Designer, and gPROMS. Nowadays, most of the current process simulators are developed following an object-oriented approach using languages such as C++ or Java. This shift in paradigm, from procedural to object-oriented, has no doubt benefited and will continue to benefit the process engineering community immensely.

Aspen Plus is designed for the simulation of steady-state processes; especially those that are computationally laborious to analyze by hand calculations, such as processes involving recycle streams, nonideal phase or chemical equilibria, and adiabatic operations. It is ideally suited to provide answers on "what-if" type of questions on process design and optimization.

Fundamental to improving performance of the plant is an accurate representation of the basic processes. Companies require a solution that enables them to model their processes to develop insights to improve designs and optimize performance. Aspen Plus provides the solution to meet this requirement, solving the critical engineering and operating problems that arise throughout the life cycle of a chemical process.

Aspen Plus predicts process behavior using engineering relationships, such as mass and energy balances, phase and chemical equilibria, and reaction kinetics. With reliable physical properties, thermodynamic data, realistic operating conditions, and rigorous equipment models, engineers are able to simulate actual plant behaviors. Applications include the following:

- Improving engineering productivity and reducing costs
- Reducing energy consumption and greenhouse gas emissions
- Enhancing product yields and quality
- Minimizing capital and operating costs
- Optimizing designs for large-scale integrated chemical plants
- Optimizing plant operations

The power and flexibility of Aspen Plus is further enhanced through a number of optional add-on applications:

- Aspen Plus Dynamics: It conducts safety and controllability studies, sizes relief valves, and optimizes transition, startup, and shutdown policies.
- Aspen Rate-based Distillation: It predicts column performance accurately over a wide range of conditions.
- Aspen Batch Modeler: Model batch reactors and columns that can be used stand-alone or inside Aspen Plus.
- Aspen Polymers: It extends Aspen Plus with a complete set of polymer thermodynamic methods and data, rate-based polymerization reaction models, and a library of industrial process models.
- Aspen Distillation Synthesis: It engages in visualization and analysis of conceptual design and troubleshooting of distillation schemes for complex mixtures.
- Aspen Energy Analyzer: It evaluates energy efficiency and optimizes heat exchanger network design.
- Aspen Custom Modeler: It develops rigorous models of special process equipment and uses them inside Aspen Plus or Aspen Plus Dynamics.

The base of Aspen Plus is flowsheet simulation, in other words, use of a computer program to quantitatively model the characteristic equations of a chemical process. Also, Aspen Plus is a good example of sequential modular approach.

4.3 Sequential Modular Simulation

The sequential modular approach is based on the concept of modularity, which extends the chemical engineering concept of unit operation to a "unit calculation" of the computer code (i.e., subroutine) responsible for the calculations of an equipment. This method is similar in principle to the traditional method of hand calculation of unit operations. The equations for each equipment unit are grouped together in a subroutine or module. Thus, each module calculates the output streams for the given input streams and parameters for that equipment, irrespective of the source of input information or the sink of output information. In the equation-oriented type, the complete model of the plant is expressed in the form of one large sparse system of nonlinear algebraic equations that is simultaneously solved for all the unknowns. This approach combines the modularizing of the equations related to specific equipment with the efficient solution algorithms for the simultaneous equation-solving technique. For each unit, an additional module is written, which approximately relates each output value by a linear combination of all input values. Accordingly, rigorous models are used at units' level, which are solved sequentially, whereas linear models are used at flowsheet level, solved globally. The linear models are updated based on results obtained with rigorous models (Martin-Martin, 2015).

Following are the basic components of a simulation package:

1. Component data bank
2. Thermodynamic property prediction methods
3. Flowsheet builder (graphical user interface)
4. Unit module library
5. Numerical routines
6. Data output generator
7. Executive program (flowsheet solver)

Modular process simulators are very robust solving each unit operation with numerical methods tailored to the specific characteristics to each one of these units. These include from specific inside-out algorithms to "flash" a material stream going through detailed methods for reactors and heat exchangers until complex methods for distillation. However, one drawback of the approach is that some unit operations introduce numerical noise (this is also simulator and unit dependent). In other words, if we solve the same problem starting from different initial points, we will obtain, for some variables, slightly different values. The difference can be in the second or third decimal point, which is not significant from a simulation point of view but is a really large error if we try to estimate a derivative (derivative information is not provided by the simulator, although some unit operations internally use it to solve the module). This problem is magnified by information loops that could act as "error accumulators."

References

Aspen Technology Inc., 2001, Aspen Plus 11.1 User Guide, Cambridge, MA, USA.

M. Martín-Martín, 2015, *Introduction to Software for Chemical Engineers*, Taylor & Francis Group: USA.

5

Direct Optimization in Aspen Plus®

5.1 Optimization Methods

Aspen Plus solves optimization problems iteratively. By default, Aspen Plus generates and sequences a convergence block for the optimization problem. It can override the convergence defaults, by entering convergence specifications on Convergence forms. It uses the sequential quadratic programming (SQP) and complex methods to converge optimization problems. The value of the manipulated variable that is provided in the Stream or Block input is used as the initial estimate. Providing a good estimate for the manipulated variable helps the optimization problem to converge in fewer iterations. This is especially important for optimization problems with a large number of manipulated variables and constraints.

There are no results associated directly with an optimization problem, except the objective function and the convergence status of the constraints. The final value of the manipulated and/or sampled variables are either reported directly on the appropriate Stream or Block results sheets or summarized on the *Results—Manipulated Variables* sheet of the convergence block. To find the summary and iteration history of the convergence block, select the Results form of the appropriate Convergence block.

The objective function can be any valid FORTRAN expression involving one or more flowsheet quantities. The tolerance of the objective function is the tolerance of the convergence block associated with the optimization problem. There is the possibility of imposing equality or inequality constraints on the optimization. Equality constraints within an optimization are similar to design specifications. The constraints can be any function of flowsheet variables computed using FORTRAN expressions or in-line FORTRAN statements. Moreover, the tolerance of the constraint must also be specified.

5.2 Sensitivity Analysis Tools in Aspen Plus

Sensitivity analysis is a tool helpful to determine how a process reacts to varying key operating and design variables. It can be used to vary one or more flowsheet variables and to study the effect of that variation on other flowsheet variables. It is a valuable tool for performing "what-if" studies, and it can be observed as a very basic method to optimize a given equipment. The flowsheet variables that are varied must be inputs to the flowsheet. They cannot be variables that are calculated during the simulation. Sensitivity analysis can be used to verify whether the solution to a design specification lies within the range of the manipulated variable. It can use sensitivity blocks to generate tables and/or plots of simulation results as functions of feed stream, block input, or other input variables. Sensitivity analysis results are reported in a table on the *Sensitivity Results Summary* sheet. The first n columns of the table list the values of the variables that are varied, where n is the number of manipulated flowsheet variables entered on the *Sensitivity Input Vary* sheet. The remaining columns in the table contain the values of variables that are tabulated on the *Tabulate* sheet.

5.3 Sequential Quadratic Programming in Aspen Plus

The SQP method is a state-of-the-art, quasi-Newton nonlinear programming algorithm. It can converge tear streams, equality constraints, and inequality constraints simultaneously using the optimization problem. It usually converges in only a few iterations, but it requires numerical derivatives for all decision and tear variables at each iteration. A description of the method can be found in the work of Gill et al. (2005). This method as implemented in Aspen Plus includes a novel feature: the tear streams can be partially converged using the Wegstein method for each optimization iteration and during line searches. This usually stabilizes convergence and can reduce the overall number of iterations. It can specify the number of Wegstein passes. Selecting a large value effectively makes SQP a feasible method (but not a black-box method). The Aspen Plus default is to perform three Wegstein passes. It can also use the SQP method as a black-box or partial black-box method, by converging tear streams and design specifications as an inside loop to the optimization problem (using separate Convergence blocks). This reduces the number of decision variables. The trade-off is the number of derivative evaluations, versus the time required per derivative evaluation. Whether SQP is the method of choice depends on the optimization problem. The default optimization convergence procedure in Aspen Plus converges tear streams and the optimization problem simultaneously using the SQP method.

5.4 Optimization of a Heat Exchanger

5.4.1 Description of the Problem

In this first case of study, a stream of ethanol is cooled, using water as cooling fluid. The process occurs in a counter-current heat exchanger, where ethanol enters the equipment at 60°C, and it is desired to obtain an output temperature of 30°C. The ethanol stream has a mass flow rate of 100 kg/h. The required flow rate of water (w_C) is unknown, and it is directly related to the change of temperature of the water stream. Nevertheless, to obtain an initial design for the exchanger, a flow rate of 100 kg/h is assumed for the cooling stream. A simplified representation of the process is shown in Figure 5.1.

5.4.2 Initial Simulation

The equipment described in Section 5.4.1 has been simulated in Aspen Plus V30.0. First, the components involved in the process (ethanol and water) are defined in the *Components—Specifications* sheet (Figure 5.2). Then, the *Next* button is pushed to move to the *Methods—Specifications* sheet, where a thermodynamic method is selected to model the phase equilibrium. This sheet can also be reached by navigating in the menu located at the left of the screen. Here, the nonrandom two-liquid (NRTL) model has been selected to represent the liquid mixture, combined with the Redlich–Kwong equation to represent the gas phase (Figure 5.3). By pushing the *Next* button, the screen shown in Figure 5.4 appears where the set of binary parameters is shown for the components of interest. In this case, all the parameters are available in the database; thus, it is possible to directly move to the next step. If the *Next* button is pushed, the simulator offers the alternative of performing a test for the properties calculations. It is highly recommendable to perform such

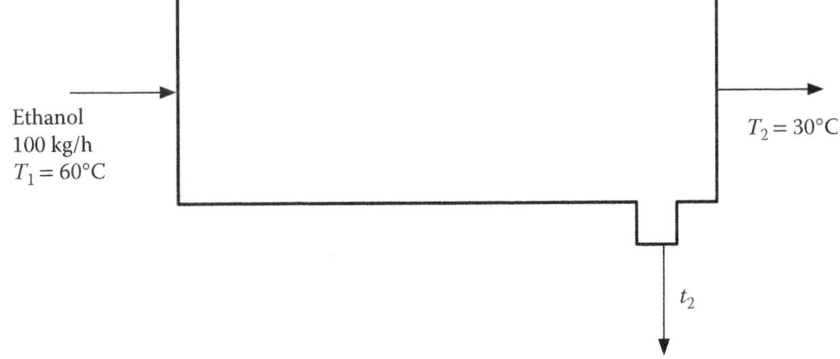

FIGURE 5.1
Heat exchanger.

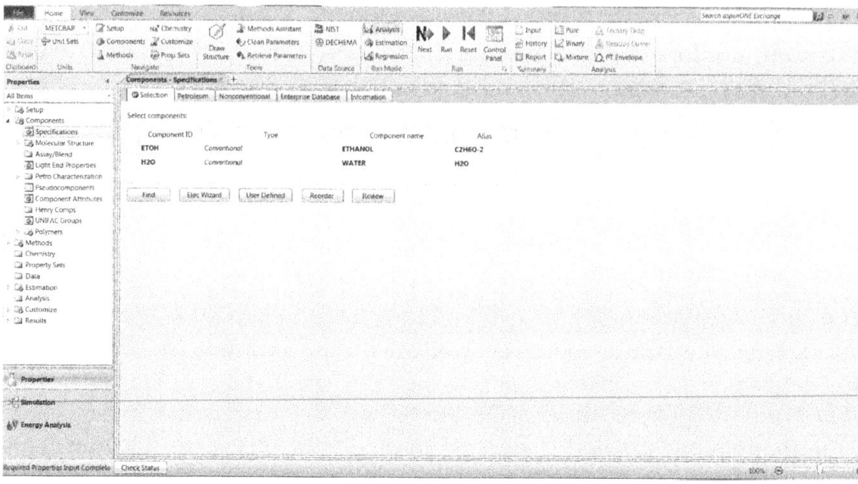

FIGURE 5.2
Selection of the components for the heat exchanger simulation.

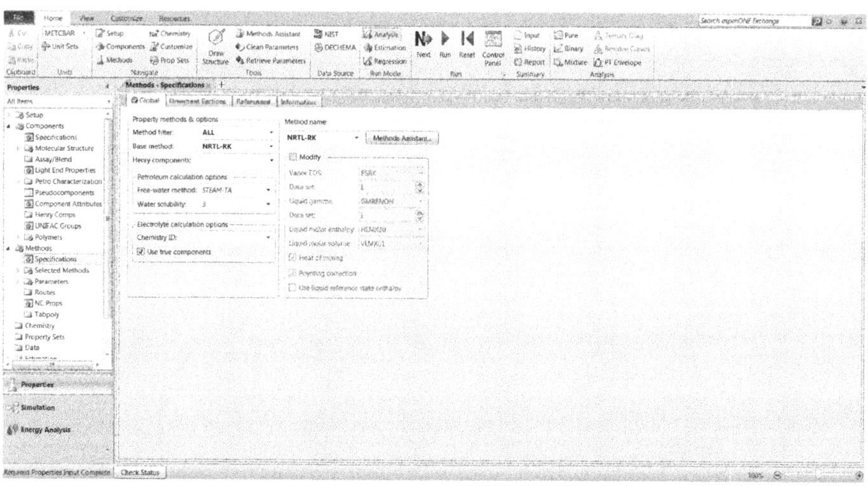

FIGURE 5.3
Selection of thermodynamic method for the heat exchanger simulation.

a test to ensure that the simulation environment has all the required data to compute the physical properties of the components of interest. Once the test for properties is performed, it is possible to move to the simulation section. In the simulation environment, a blank flowsheet appears. Thus, it is necessary to insert the required process equipment. Aspen Plus includes a library with different process equipment, such as mixers, separators, and columns.

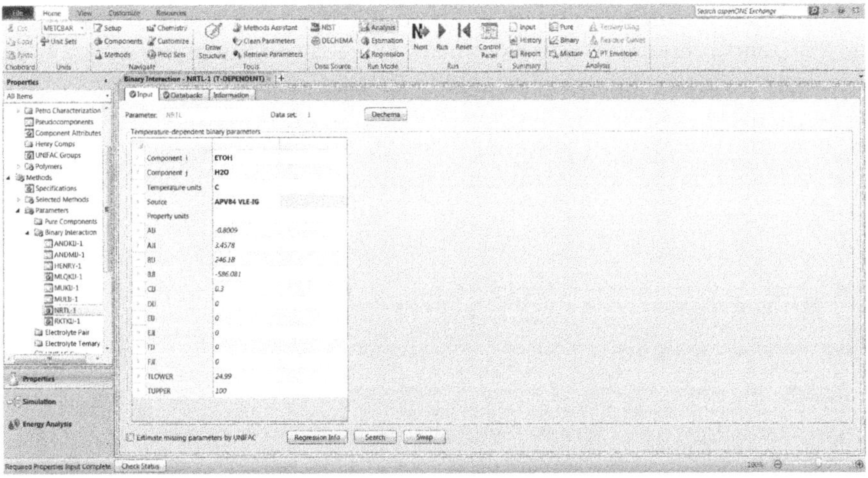

FIGURE 5.4
Binary interaction parameters for the ethanol–water mixture.

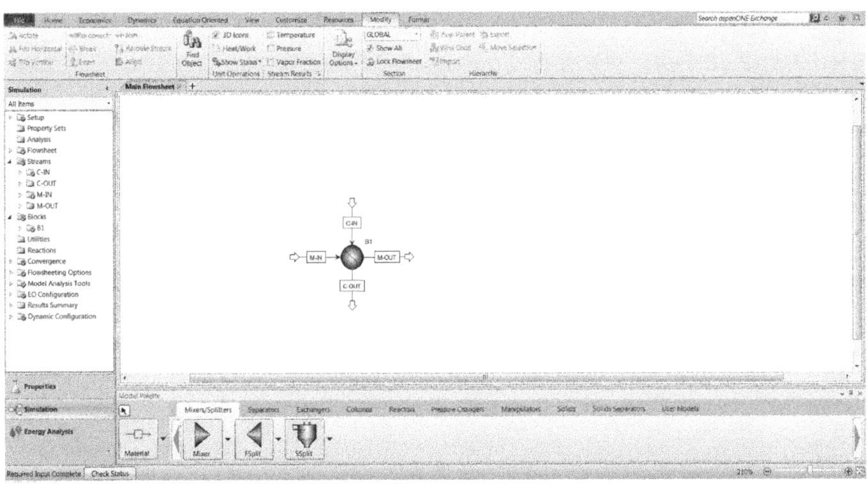

FIGURE 5.5
Flowsheet for the heat exchanger simulation.

In this case, we are interested in a heat exchanger; thus, the *Exchangers* menu is selected. Four types of exchangers can be observed in the menu. The *HeatX* alternative is selected because it allows simulating an exchanger with two fluid streams. The module is inserted in the flowsheet and material streams are added (Figure 5.5). It is now necessary to define the state of the streams entering into the exchanger. The water stream, as aforementioned, is assumed to enter with a mass flow rate of 100 kg/h. Temperature is defined as 25°C and pressure as 1 bar (Figure 5.6). Feed conditions for ethanol are similarly defined (Figure 5.7). Once the streams entering the equipment are defined, the characteristics of the exchanger must be indicated. For this case, a shortcut calculation is selected as design approach. The specification for the exchanger is the hot stream outlet temperature, which is defined as 30°C

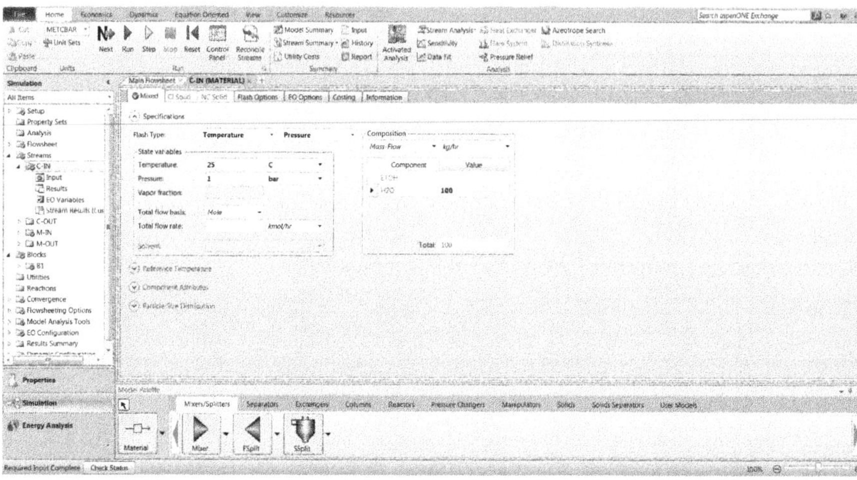

FIGURE 5.6
Input data for the cold stream of the heat exchanger.

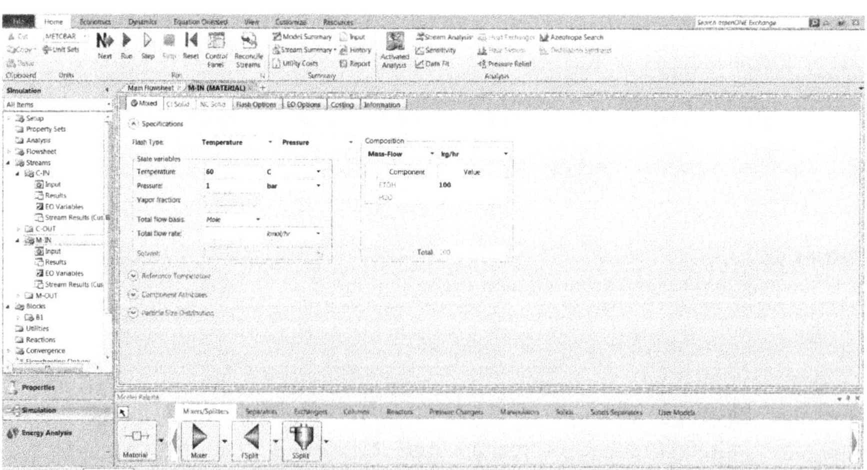

FIGURE 5.7
Input data for the hot stream of the heat exchanger.

(Figure 5.8). Once all the required information has been uploaded, it is necessary to run the simulation. This can be done by pushing the *Next* button or directly the *Run* button. If the message *Results Available* appears, it indicates that all the calculations have been successfully completed, and the results are reliable. For the heat exchanger, the summary of the thermal results are shown in Figure 5.9. In the results, it can be observed that both streams remain as liquid and that the water stream leaves the exchanger at 45.03°C.

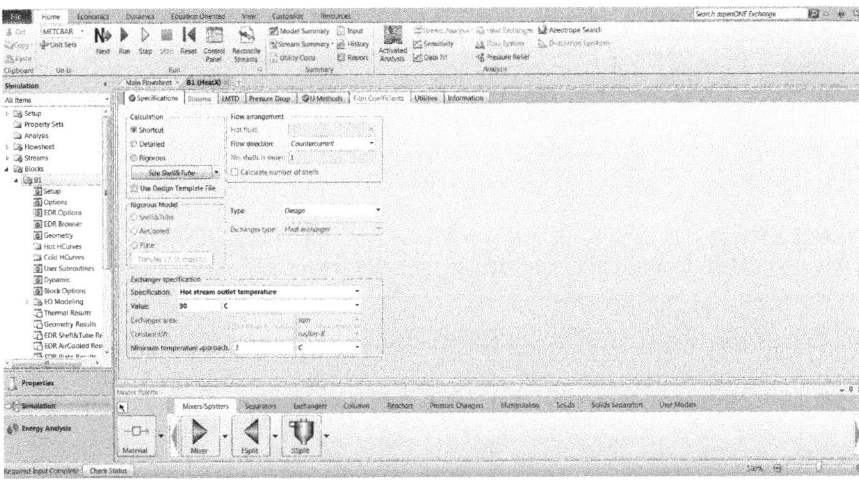

FIGURE 5.8
Specifications sheet for the heat exchanger.

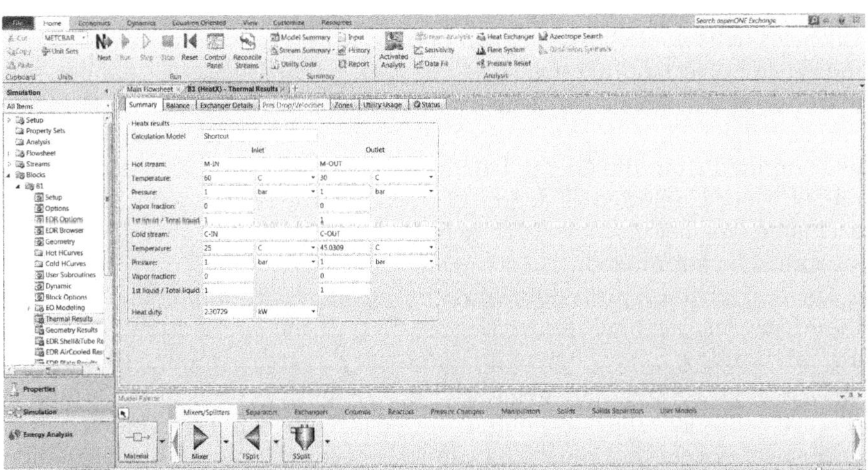

FIGURE 5.9
Results summary for the initial simulation of the heat exchanger.

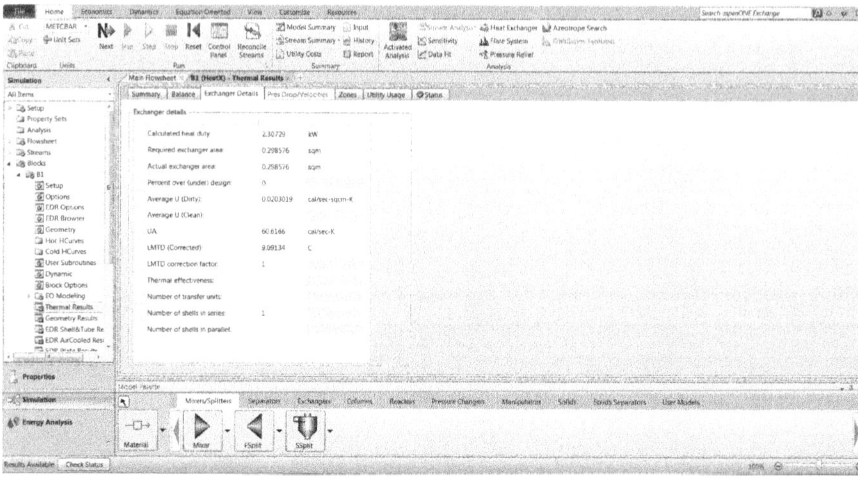

FIGURE 5.10
Exchanger details for the initial simulation of the heat exchanger.

The heat duty of the exchanger is 2.31 kW. Figure 5.10 shows more detailed results for the exchanger, where it is observed that, for the case under analysis, an exchanger area of 0.298 m^2 is necessary.

5.4.3 Optimization through Sensitivity Analysis

Once the initial simulation for the exchanger is successfully obtained, the optimization procedure takes place. As a first approach, a sensitivity analysis has to be applied. To do such analysis, it is necessary to determine which variables will be selected as degrees of freedom. It is also necessary to decide the type of objective function to be minimized. For the heat exchanger, two main variables can act as a degree of freedom: the cooling stream flow rate and its outlet temperature. If one of these variables is fixed, the other one can be computed from the energy balances. For illustration purposes, the flow rate of water is selected as degree of freedom. The objective function is the exchange area because it is directly related to the equipment cost. Thus, the area must be minimized.

The optimization procedure starts in the *Sensitivity* subfolder, located in the *Model Analysis Tools* folder (Figure 5.11). If the *New* button is pushed, a small window opens, where it is necessary to create a name (ID) for the optimization routine. The default ID is S-1, and it is used in this example (Figure 5.12). Once the name of the routine is defined, the simulation moves to a set of sheets that must be completed with the required information (Figure 5.13). In the *Vary* sheet, the variables to be manipulated are loaded, using the *New* button to create manipulated variables. Once the manipulated variable is created, its type is set as *Stream-Var*, because the degree

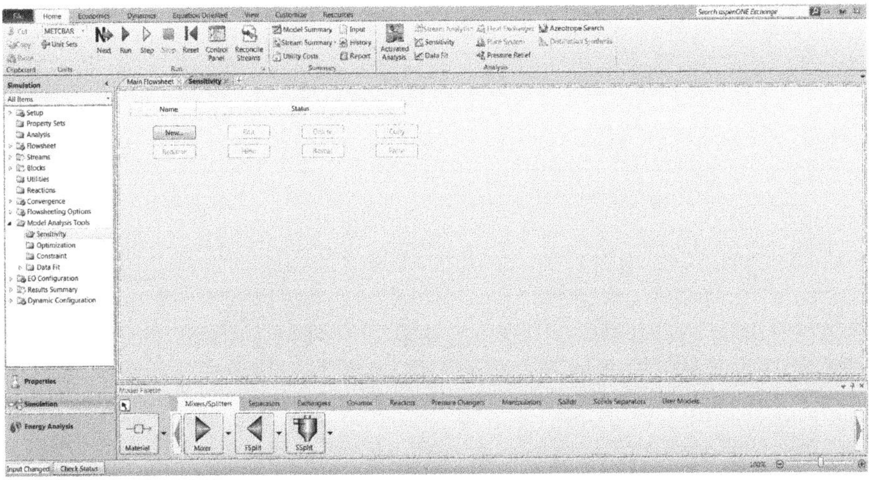

FIGURE 5.11
Sensitivity subfolder.

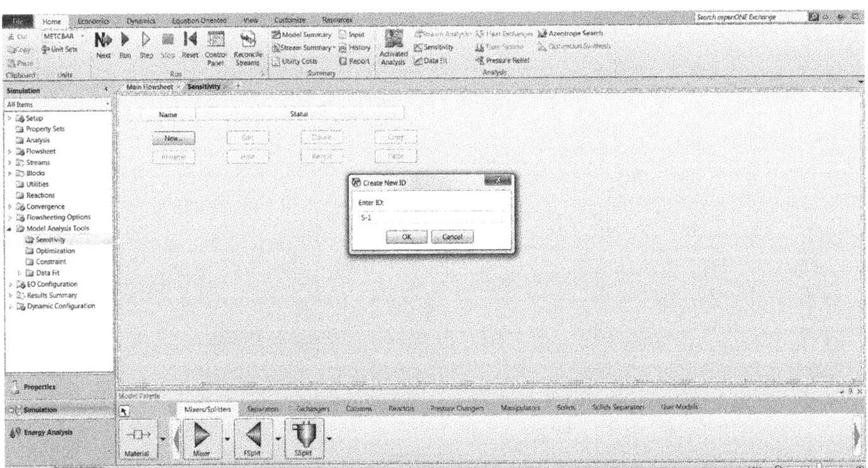

FIGURE 5.12
Creation of a new sensitivity analysis for the heat exchanger.

of freedom in this case is a mass flow rate, which is a stream variable. The name of the stream is also defined as *C-IN* (which is the name of the water stream, according to Figure 5.5) and the type of variable is selected as *MASS-FLOW*. The limits for the manipulated variable are set as 50 and 550 kg/h, respectively, to allow a wide search space. Finally, the number of points to be analyzed is required. For this example, 100 points are selected. The *Vary* sheet with all the information is shown in Figure 5.14. Now, it is necessary to define the measured variables, which are provided by the

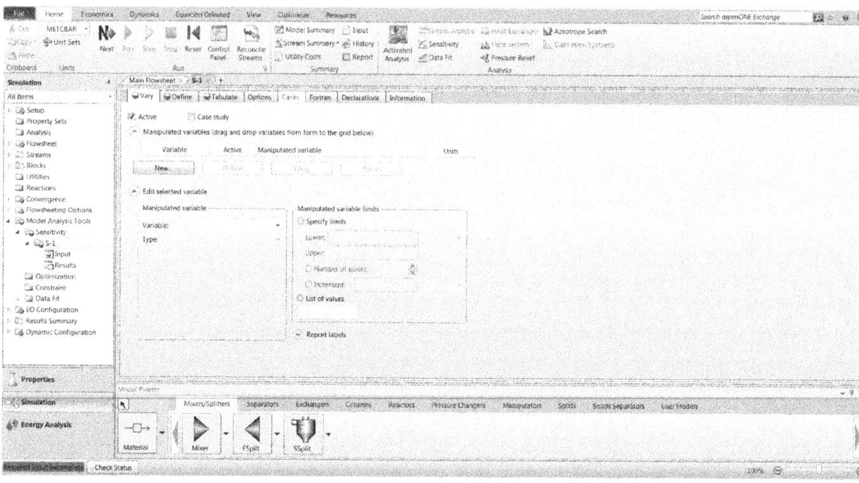

FIGURE 5.13
Vary sheet for the sensitivity analysis.

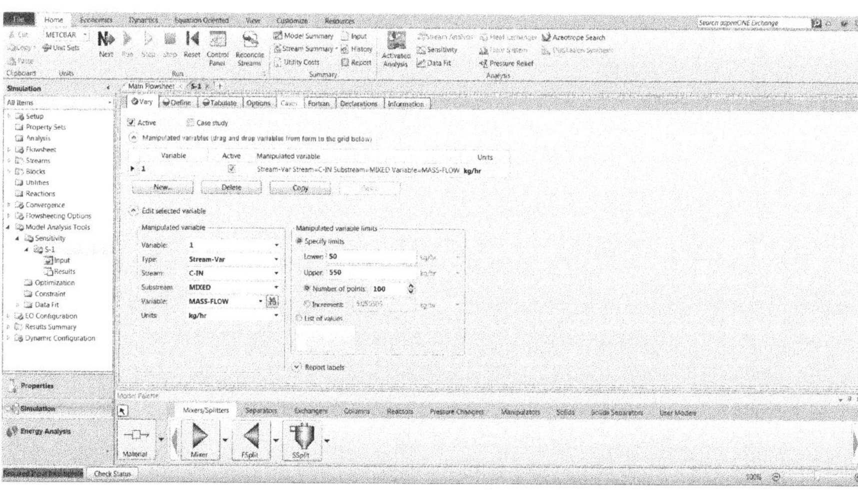

FIGURE 5.14
Completed *Vary* sheet for the sensitivity analysis of the heat exchanger.

selected objective function, i.e., the exchanger area. This can be done in the *Define* sheet. The user identifier of the variable is defined in the *Variable* field. In this case, the variable is identified as *AREA*. Then, the category of the variable is selected. Because the area is a design variable for the exchanger block, its category is established as *Blocks*. The name of the variable area used by the simulation engine is *AREA-CALC*; thus, that name is selected in the *Variable* menu. The *Define* sheet with all the required information loaded is shown in Figure 5.15. Finally, in the *Tabulate* sheet, it is necessary

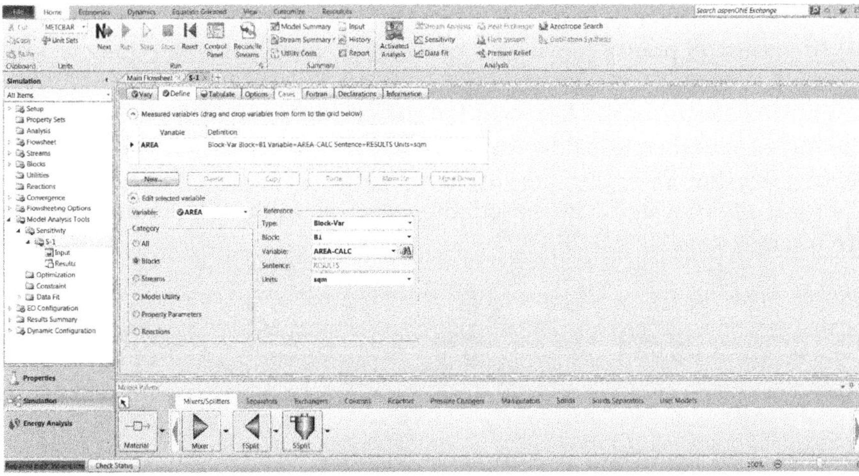

FIGURE 5.15
Define sheet for the sensitivity analysis of the heat exchanger.

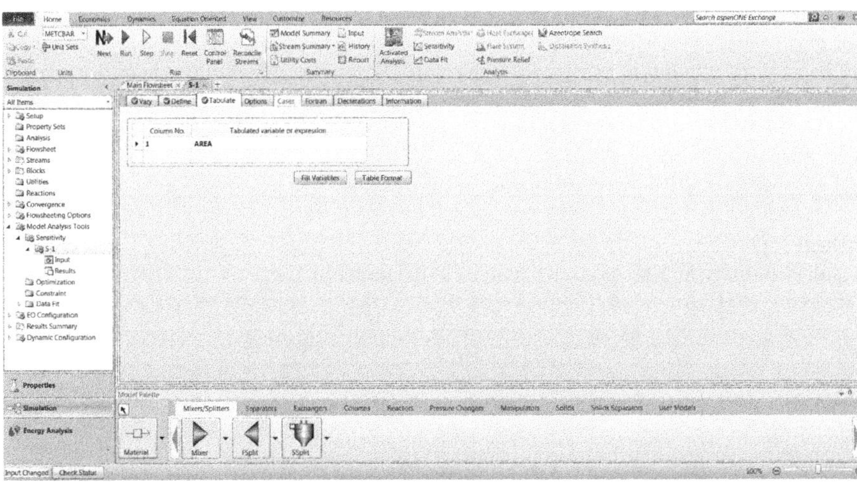

FIGURE 5.16
Tabulate sheet for the sensitivity analysis of the heat exchanger.

to indicate which of the variables must be shown as part of the results. If all the variables on the *Define* sheet are to be tabulated, the *Fill Variables* button can be used to complete the form. In this case, the only variable to be tabulated is *AREA*, as shown in Figure 5.16. Once all the required data have been loaded, the *Next* button (or the *Run* button) is pushed to run the simulation. The *Results* sheet is shown in Figure 5.17. It can be observed that an error message appears at the left bottom of the screen. This is due to a temperature crossing in the first two points analyzed. Nevertheless, all the other 98 points converge to a solution; thus, the first two cases are ignored. A plot of the exchanger area against the water flow rate is shown in Figure 5.18. It can be observed that the exchanger area diminishes with the increase in the mass flow rate of water. Nevertheless, for values of flow rate higher than 200 kg/h, the variation in area is small. As expected, when the flow rate of water is increased, the variation in the temperature of that stream gets smaller.

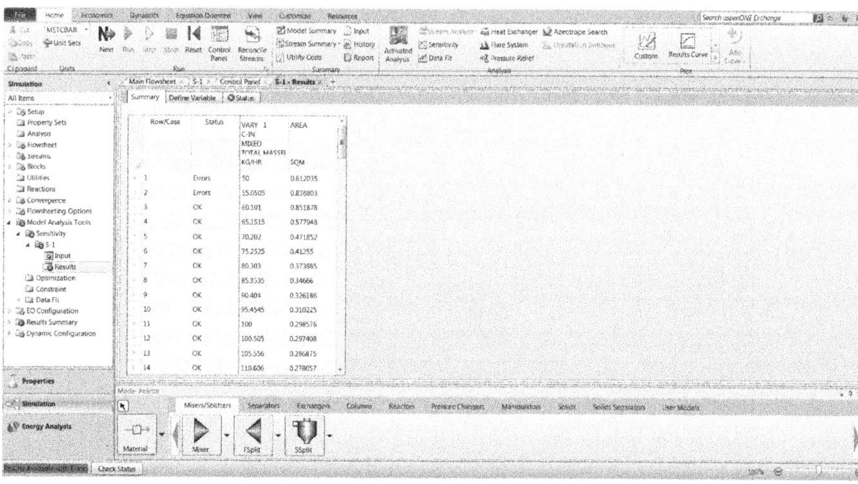

FIGURE 5.17
Results summary for the sensitivity analysis of the heat exchanger.

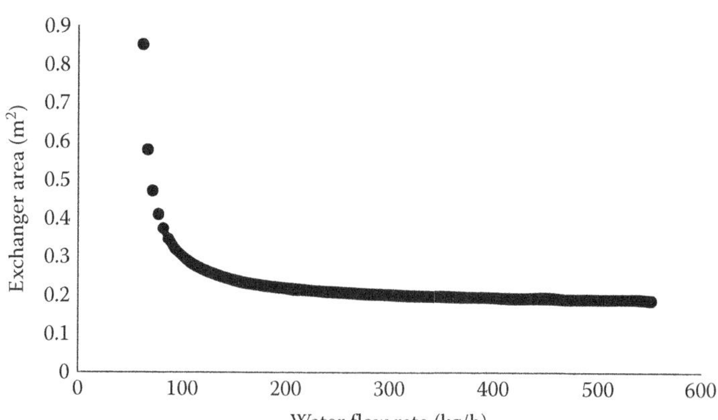

FIGURE 5.18
Exchanger area as a function of water flow rate.

5.4.4 Optimization through Sequential Quadratic Programming

In this case, the optimization procedure is performed using the SQP method, which is the default method for optimization. This can be observed in the *Options* subfolder of the *Convergence* folder, as shown in Figure 5.19. To start the optimization procedure, the *Optimization* subfolder, located inside the *Model Analysis Tools* folder, must be opened. Here, the *New* button is pushed to open a new optimization subfolder, as shown in Figure 5.20. Then, a name

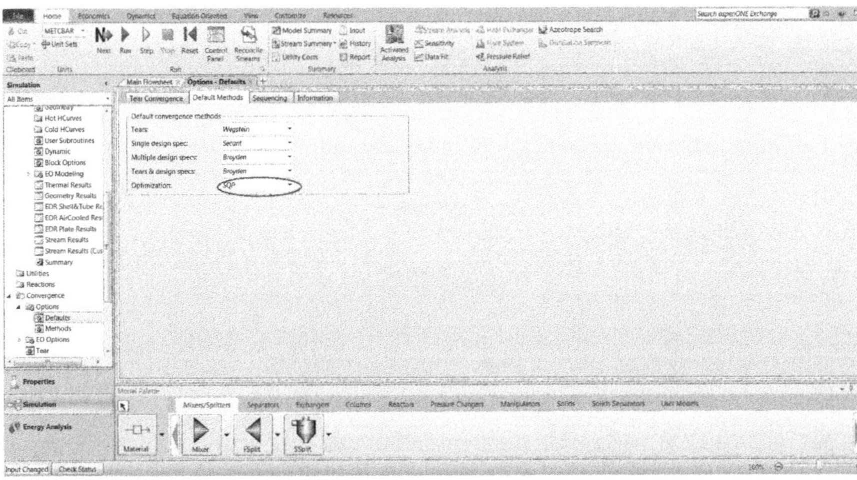

FIGURE 5.19
Checking the optimization algorithm.

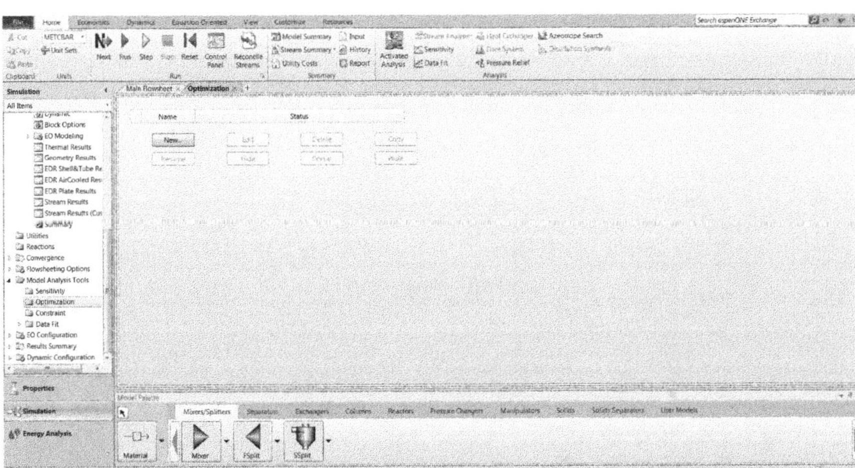

FIGURE 5.20
Optimization subfolder.

(ID) is created to represent that folder. The default ID can be used (Figure 5.21). Once the optimization subfolder is created, it is necessary to define the variables to be measured. First, a name for each variable must be written in the *Variable field*. In Figure 5.22, a variable named *AREA* is defined. This name is only a tag for the management of variables by the user. Nevertheless, once the name of the variable is created, it must be linked with the variables of the simulator. In this case, the area is a block variable, thus that category is selected. The tag of the block to which the variable corresponds must be

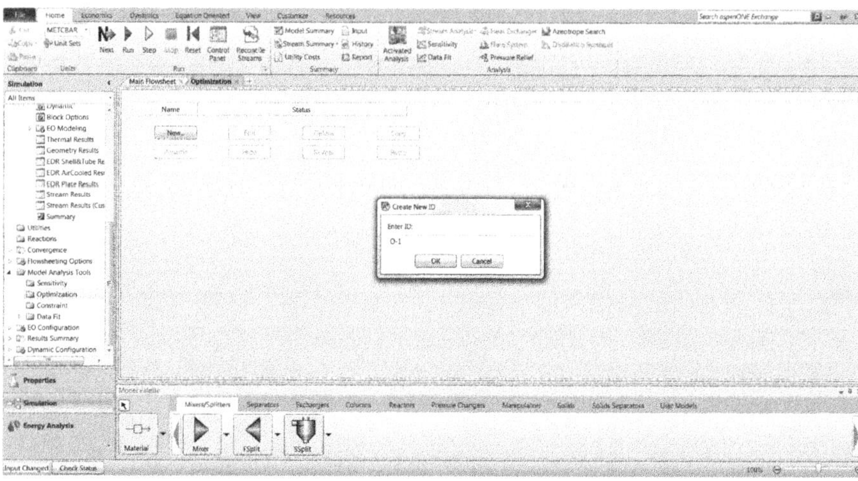

FIGURE 5.21
Creation of a new optimization procedure for the heat exchanger.

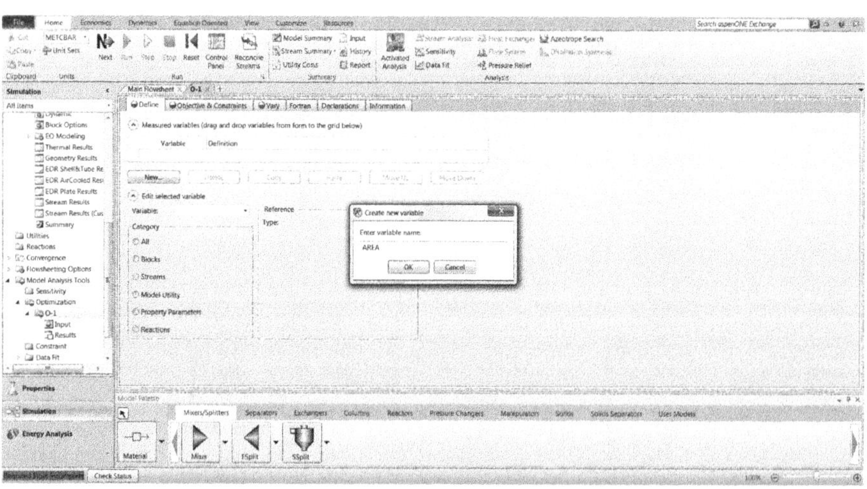

FIGURE 5.22
Creation of a new measured variable for the heat exchanger.

selected in the *Block* menu. Now, the simulated variable must be selected from a list of variables of that block. In this case, the variable *AREA-CALC* is selected, because it represents the calculated area of the exchanger. All this information is loaded in the *Define* sheet, as presented in Figure 5.23. The other variables involved in the optimization procedure are the outlet temperature of the cold stream and the water flow rate. The water flow rate is the selected degree of freedom. The temperature is defined to obtain the information about that variable in the final report, although it is not properly a degree of freedom. Other variables could be added if they are of interest. The definition of these two variables can be observed in Figures 5.24 and 5.25. Once all the variables are defined, the objective function and the constraints are loaded in the *Objective & Constraints* sheet. Here, the variable *AREA* is loaded as the objective to be minimized. In this case, no constraints are added (Figure 5.26). Finally, the variables to be manipulated to reach the optimal solution (i.e., the degrees of freedom) are defined in the *Vary* sheet. In this case, the mass flow rate of the water stream (C-IN) is established as the manipulated variable, with a range from 50 to 550 kg/h (Figure 5.27). Once all the data have been loaded, the *Next* button is pushed to run the simulation. If the *Results Available* message appears, it indicates that an optimal solution is reached. Figure 5.28 shows the solution, which can be observed in the *Results* menu of the O-1 optimization subfolder. It can be observed that, if a mass flow rate of 550 kg/h of water is used, the area is reduced to 0.189 m^2, with an outlet temperature of 28.67°C for the water. Nevertheless, it must be recalled from the sensitivity analysis that the area does not show a defined minimum, but it is slightly reduced as the water flow rate is increased. Thus, the solution obtained by the SQP algorithm cannot be considered a minimum from a rigorous point of view.

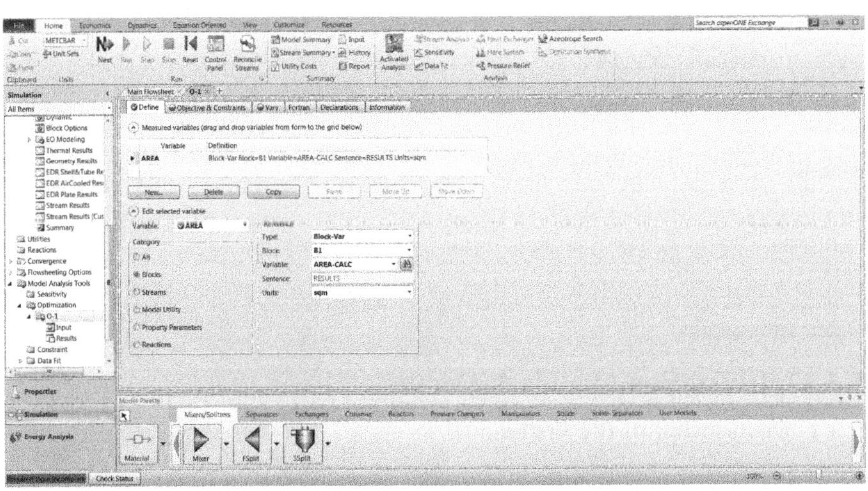

FIGURE 5.23
Define sheet for the area of the heat exchanger.

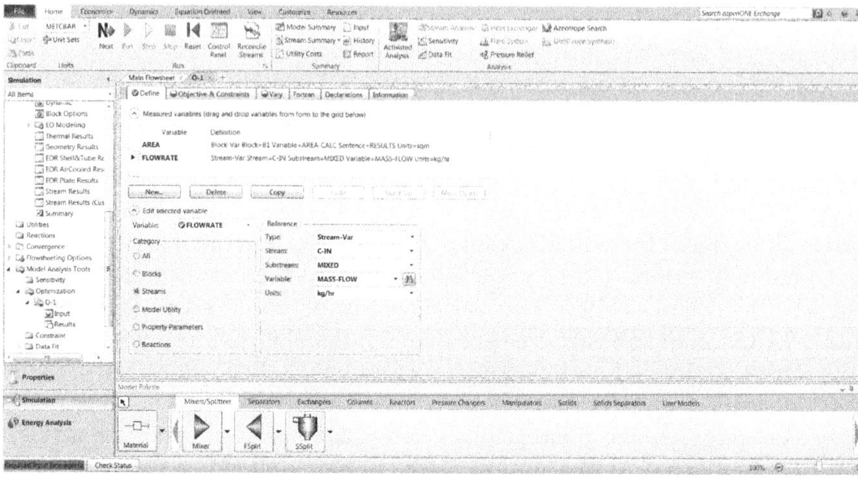

FIGURE 5.24
Define sheet for the mass flow rate of water entering the heat exchanger.

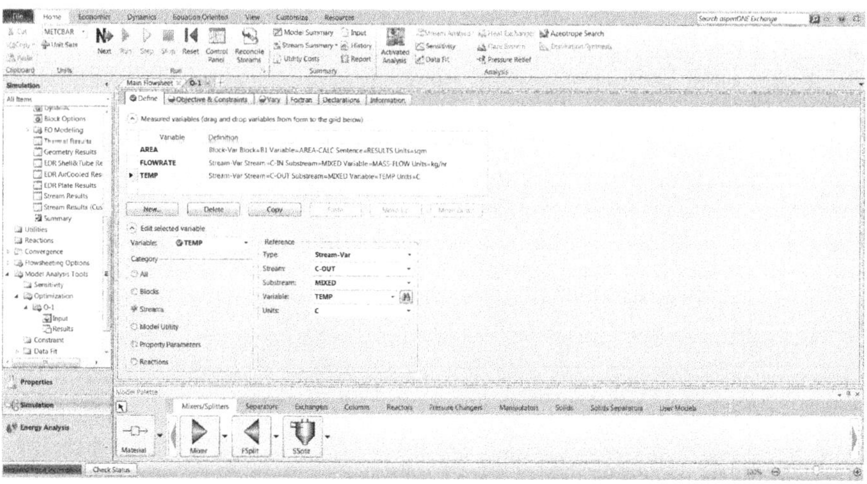

FIGURE 5.25
Define sheet for the temperature of the stream leaving the heat exchanger.

5.5 Optimization of a Flash Drum

5.5.1 Description of the Problem

For the second case of study, the partial separation of a binary mixture by vaporization is studied. The operation occurs in a flash drum. The feed stream is a mixture of methanol (A) and water (B), with composition of 60 mol% of A.

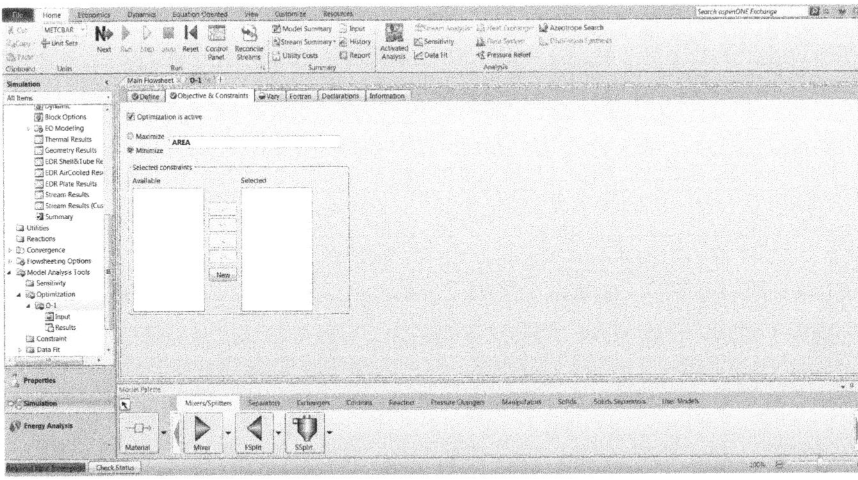

FIGURE 5.26
Objective function for the heat exchanger.

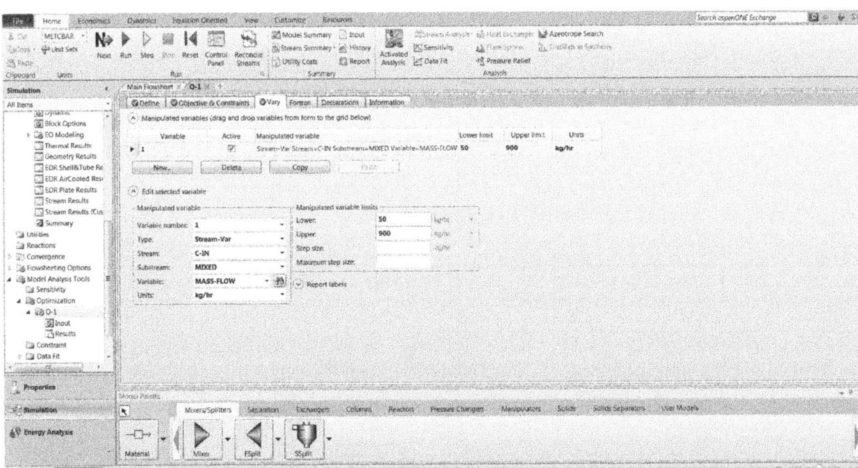

FIGURE 5.27
Vary sheet for the heat exchanger.

A mass flow rate of 100 kmol/h is processed. It is desired to obtain as much A as possible in the vapor phase, with a maximum purity of 20 mol% of the heavy component in the same stream. It is assumed that the feed stream enters into the flash drum at a pressure of 1 bar and at a temperature of 25°C, and that the operation pressure of the drum is 1 bar. Figure 5.29 shows a representation of the process.

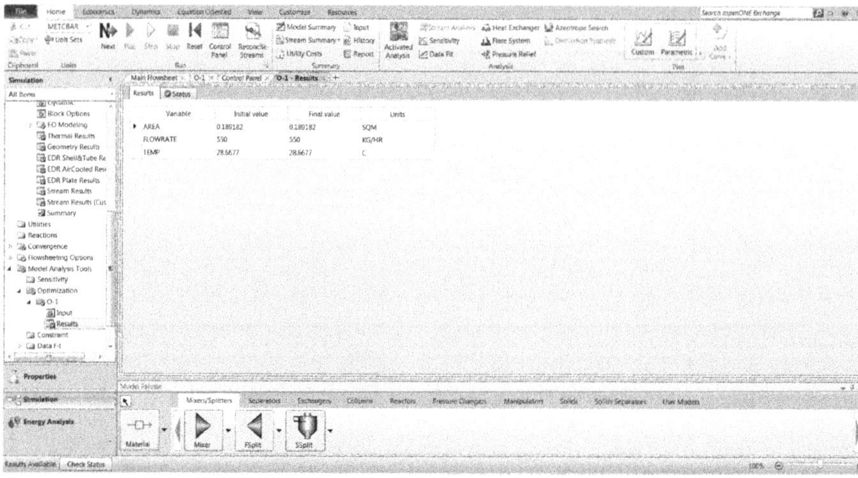

FIGURE 5.28
Results for the optimization of the heat exchanger.

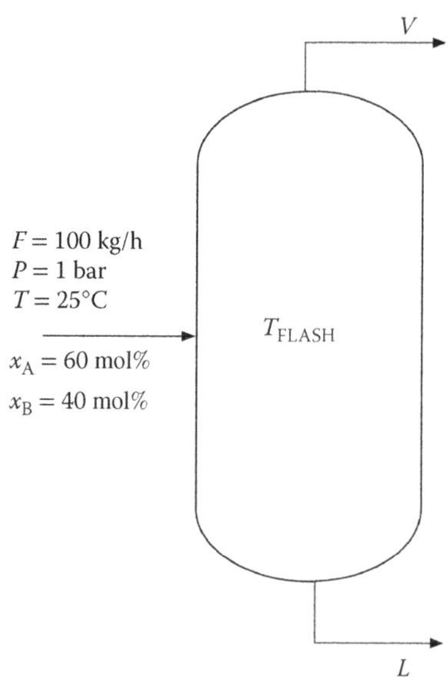

FIGURE 5.29
Flash drum.

5.5.2 Initial Simulation

The initial simulation of the flash drum is described in this section. The two components involved in the process (methanol and water) are first defined in the *Components—Specifications* sheet, as shown in Figure 5.30. Then, in the *Methods—Specifications* sheet, the NRTL model is selected to represent the

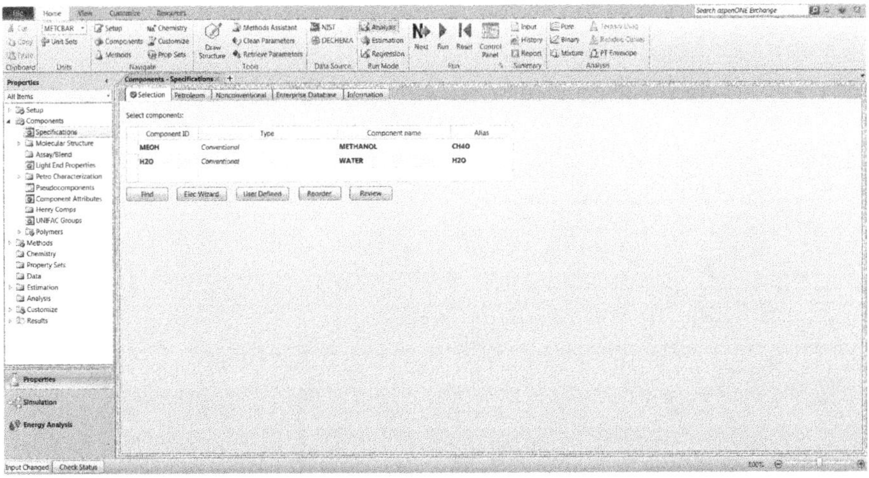

FIGURE 5.30
Selection of the components for the flash drum simulation.

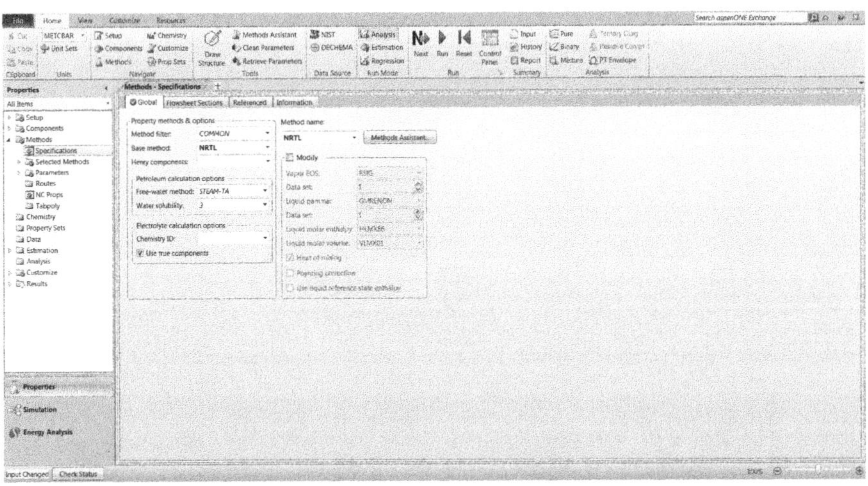

FIGURE 5.31
Selection of thermodynamic method for the flash drum simulation.

vapor–liquid equilibrium (Figure 5.31). The binary interaction parameters obtained from the Aspen database are used for the simulation (Figure 5.32). With this information, it is possible to run the properties test. If this test provides positive results, it is possible to move toward the simulation section, where the flash drum is inserted (Figure 5.33). The equipment can be found in the *Separators* menu as a *Flash2* module, which is selected because the required separation involves only two phases (liquid and vapor). The material streams, exhibiting a single feed stream and two product streams,

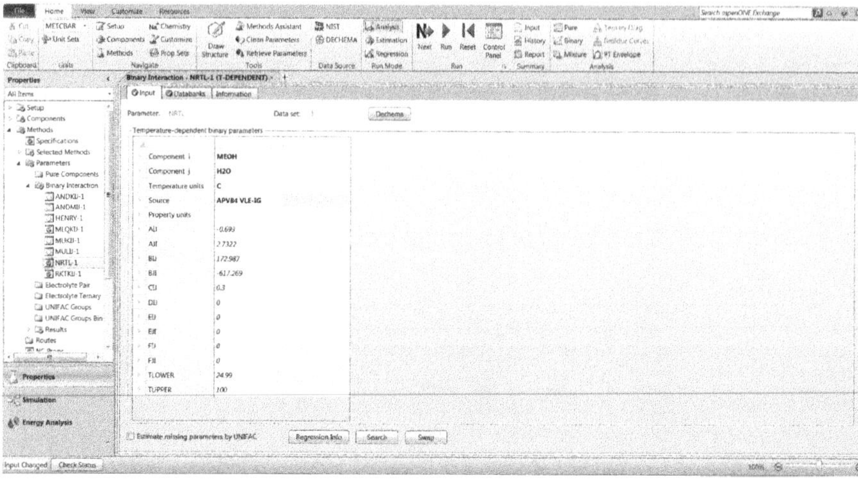

FIGURE 5.32
Binary interaction parameters for the methanol–water mixture.

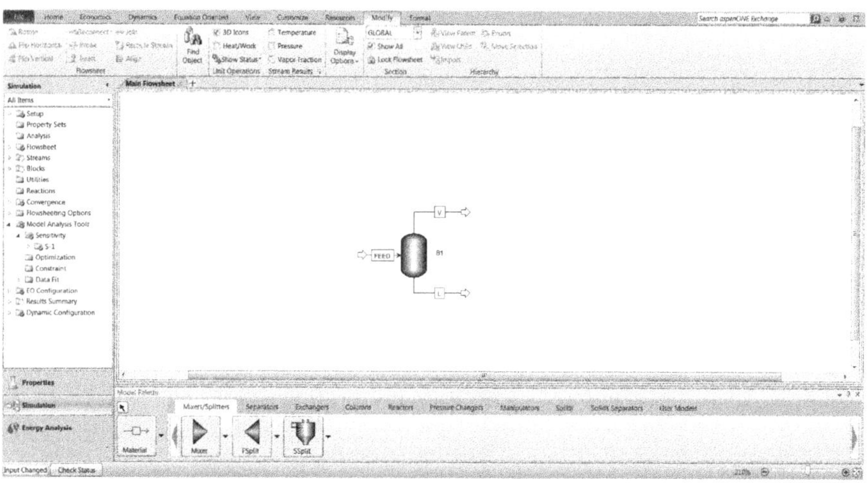

FIGURE 5.33
Flowsheet for the flash drum simulation.

are then added. Now, the conditions of the feed stream are defined. As mentioned in Section 5.5.1, it is assumed that the feed stream enters the equipment at 1 bar and 25°C. The feed flow rate and composition have also been defined previously, thus it is directly loaded in the *Input* sheet (Figure 5.34). Now, the characteristics of the flash must be loaded. Two specifications are required, which by default are the pressure and temperature of the flash. That is the combination of variables to be used in this example. The pressure

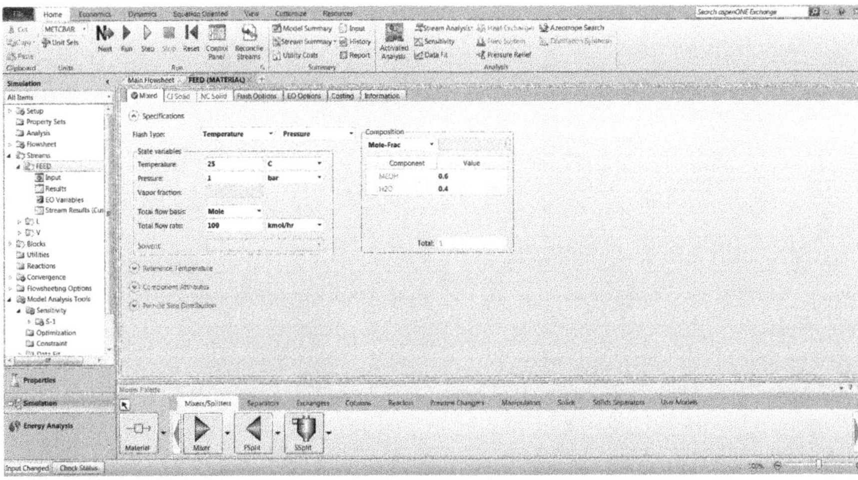

FIGURE 5.34
Input data for the stream entering the flash drum.

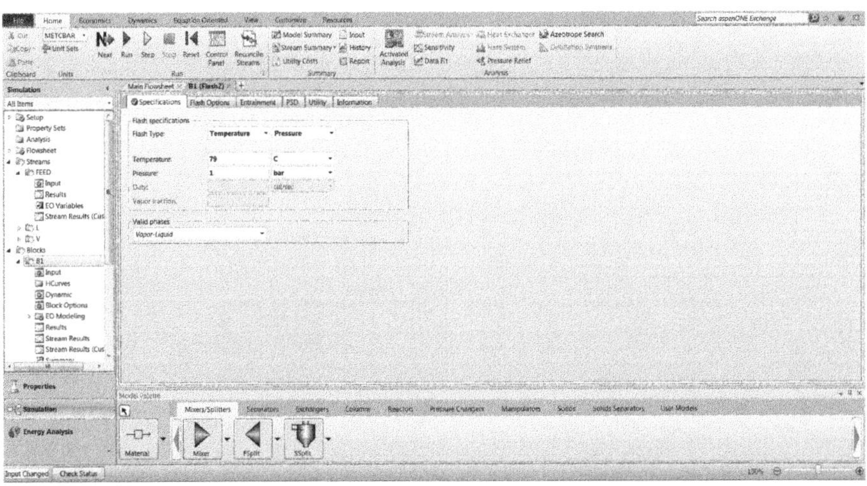

FIGURE 5.35
Specifications sheet for the flash drum.

is set as 1 bar. The required temperature is not known, but a weighed average between the boiling temperature of methanol (64.7°C) and water (100°C) can be used as a first guess (Figure 5.35). This is all the information required for the flash drum. Thus, it is possible now to run the simulation. Once the calculations are done, it is possible to revise the *Results* sheet, where the vapor fraction and heat duty are observed (Figure 5.36). Nevertheless, at this stage, we are more interested in the *Stream Results* sheet, because it reveals whether the separation task has been accomplished. From that

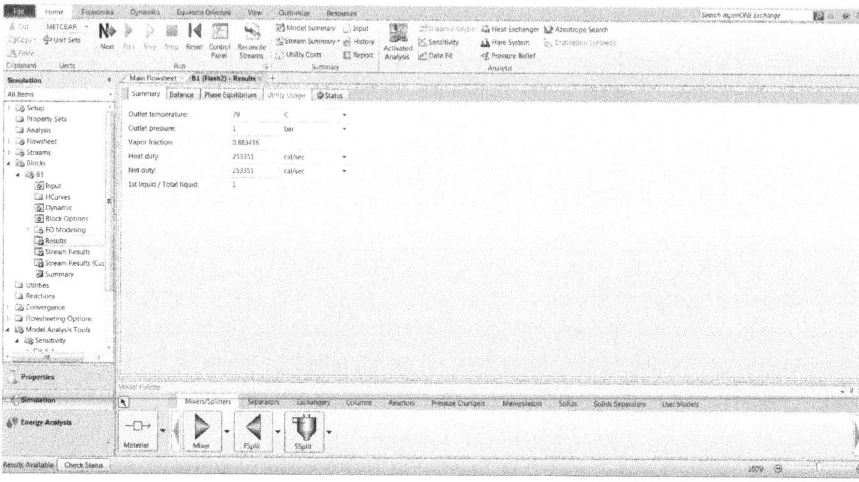

FIGURE 5.36
Results summary for the initial simulation of the flash drum.

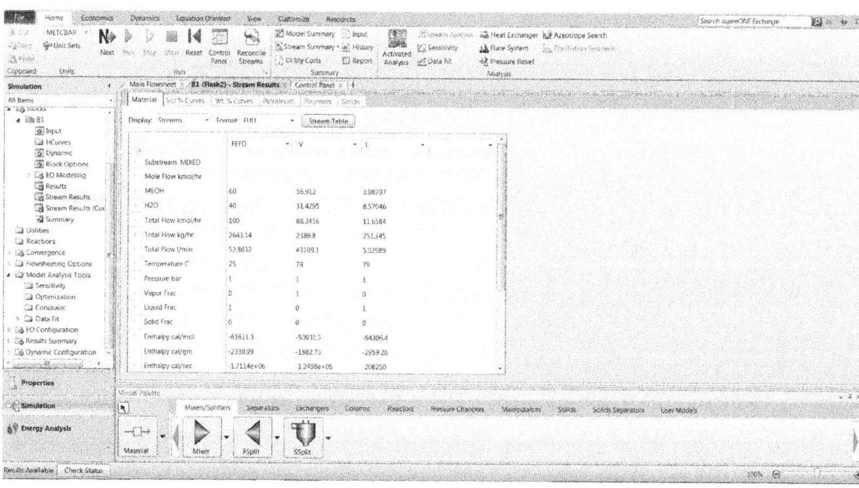

FIGURE 5.37
Streams results for the initial simulation of the flash drum.

sheet (Figure 5.37) it can be computed that, although the molar flow rate of methanol in the vapor phase is high, the purity of water in the same phase is approximately 35.5 mol%, which is higher than 20 mol% established as maximum. This implies that the assumed temperature was considerably high. The temperature could be manipulated to reach the desired purity of water. Nevertheless, this would be automatically performed by using the sensitivity analysis.

5.5.3 Optimization through Sensitivity Analysis

As aforementioned, the main manipulated variable for the flash drum is the operation temperature. The objective is to maximize the molar flow rate of methanol in the vapor stream. In this case, the maximum molar composition of water in the vapor stream acts as an inequality constraint. Because it is not possible to include constraints in the sensitivity analysis, the molar composition of water will be computed and the cases that accomplish the constraint will be considered as the candidates for the optimal solution.

To initiate the analysis, a new ID is created in the *Sensitivity* subfolder of the *Model Analysis Tools* folder. The default ID S-1 is used in this case (Figure 5.38). Now the information required in the *Vary* sheet must be completed. The manipulated variable is considered as the temperature of the flash, which is a block variable (*Block-Var*) defined as *TEMP*. The limits for the manipulated variable are set as 60°C and 100°C. This range comprises the boiling temperatures of the pure components. The number of points to be analyzed is set as 100. All the mentioned information is shown in Figure 5.39. Next, the measured variables are defined. In this case, the objective function (the molar flow rate of methanol in the vapor stream) is the main measured variable. That flow rate is in the *Streams* category, with a type *Mole-Flow* (Figure 5.40). Nevertheless, the molar composition of water in the vapor stream is also defined as measured variable to evaluate the potential solutions that accomplish the inequality constraint. The molar composition is also in the category *Streams*, but its type is defined as *Mole-Frac* (Figure 5.41). In the *Tabulate* sheet, the *Fill Variables* button is pushed to indicate that both variables of the *Define* sheet are tabulated (Figure 5.42). Now, the *Next* button (or the *Run* button) is pushed to run the simulation. Figure 5.43 shows the *Results* sheet. Methanol

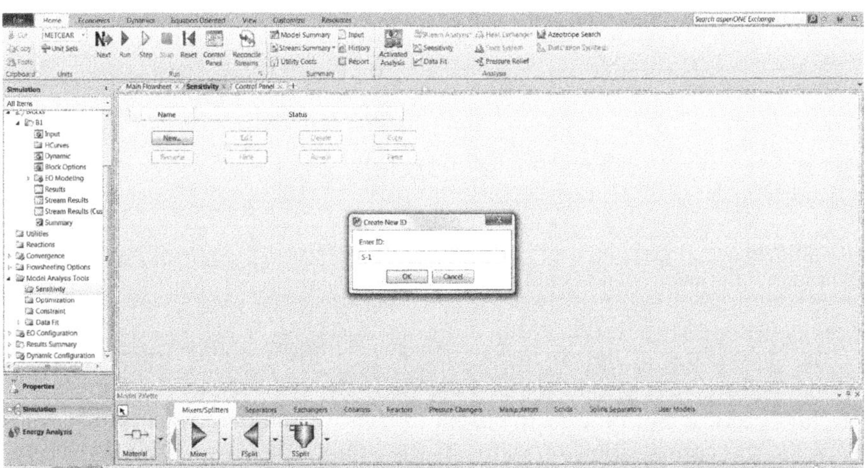

FIGURE 5.38
Creation of a new sensitivity analysis for the flash drum.

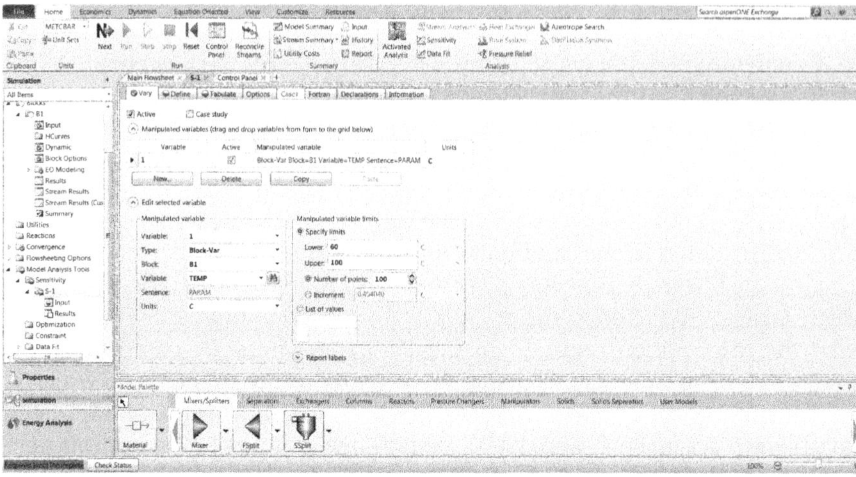

FIGURE 5.39
Completed *Vary* sheet for the sensitivity analysis of the flash drum.

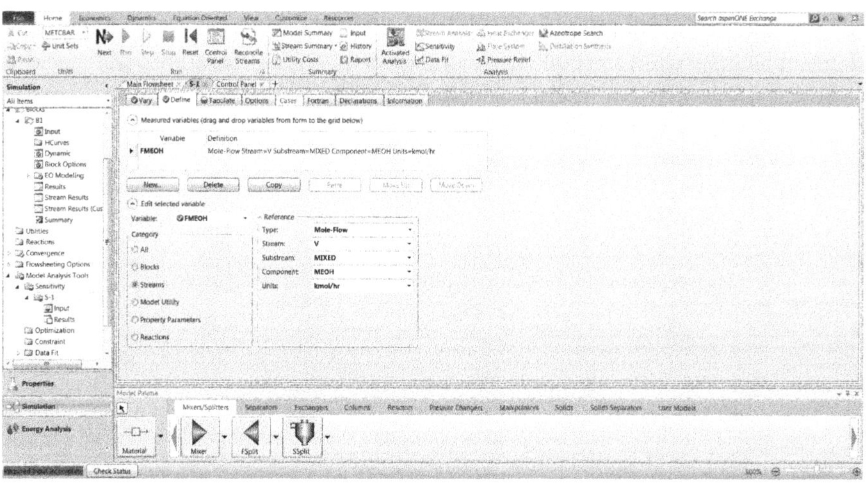

FIGURE 5.40
Definition of the methanol flow rate as measured variable.

starts vaporizing at approximately 70.90°C, with a molar composition of water in the vapor stream of 16.6 mol%. If the data shown in Figure 5.43 are analyzed, it can be clearly observed that for temperatures higher than 72.52°C, the composition of water in the vapor stream exceeds the upper bound of 20 mol%. Thus, the maximum flow rate of methanol, which can be obtained in the vapor stream without violating the inequality constraint, is 19.6842 kmol/h at 72.12°C.

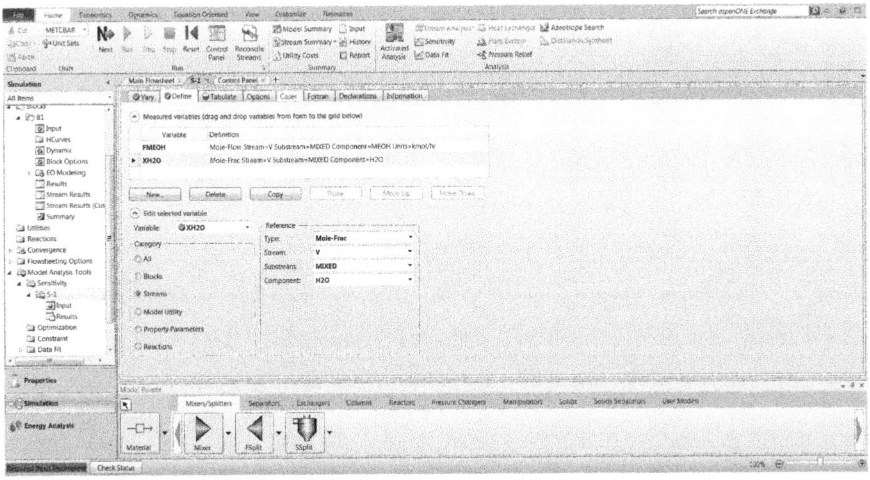

FIGURE 5.41
Definition of the water mole fraction as measured variable.

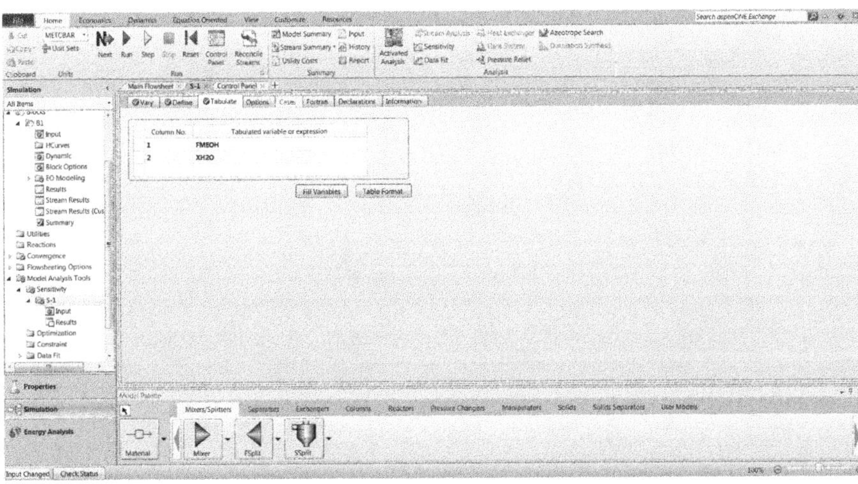

FIGURE 5.42
Tabulate sheet for the sensitivity analysis of the flash drum.

5.5.4 Optimization through Sequential Quadratic Programming

In this section, the optimization of the flash drum using the SQP method is presented. First, a new optimization routine is created in the *Optimization* subfolder of the *Model Analysis Tools* folder. The optimization routine is identified with the ID O-1, as shown in Figure 5.44. Now, the measured variables are defined. For our case of study, the variables involved in the optimization procedure are the following: the flash temperature, the molar

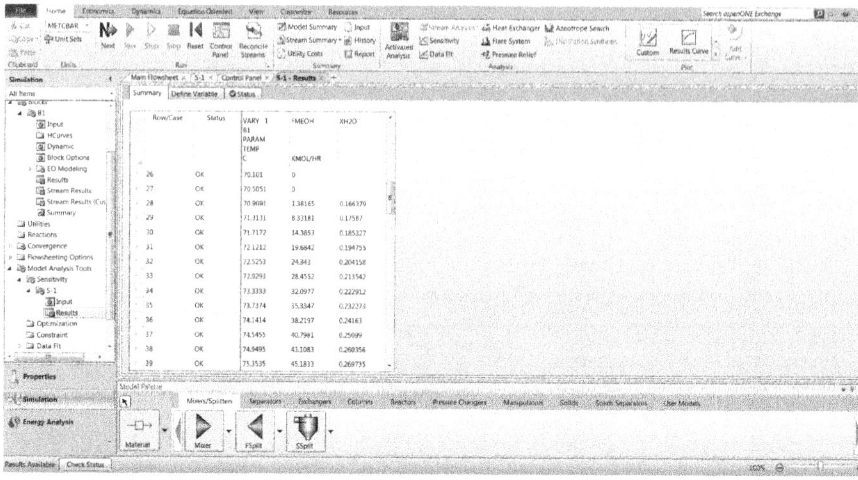

FIGURE 5.43
Results summary for the sensitivity analysis of the flash drum.

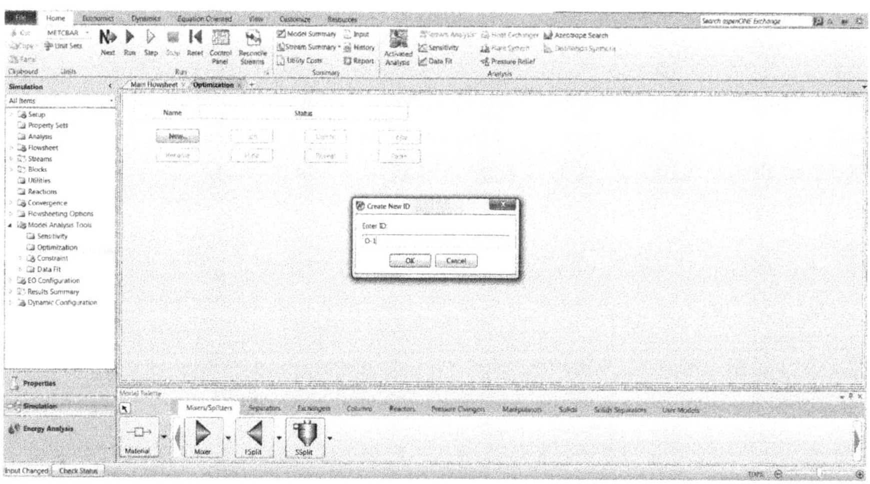

FIGURE 5.44
Creation of a new optimization procedure for the flash drum.

flow rate of methanol in the vapor stream, and the molar composition of water in the same stream. The temperature is first defined, using the identifier *TEMP*. This variable is characterized in the *Blocks* category, with the type *Block-Var* (Figure 5.45). The molar flow rate of methanol is identified as *FMEOH*, falling in the category *Streams*, with a type *Mole-Flow* (Figure 5.46). Finally, mole composition of water is identified as X_{H_2O}, being in the *Streams* category, with a type *Mole-Frac* (Figure 5.47). Now, the objective function and the constraints are loaded in the *Objective & Constraints*

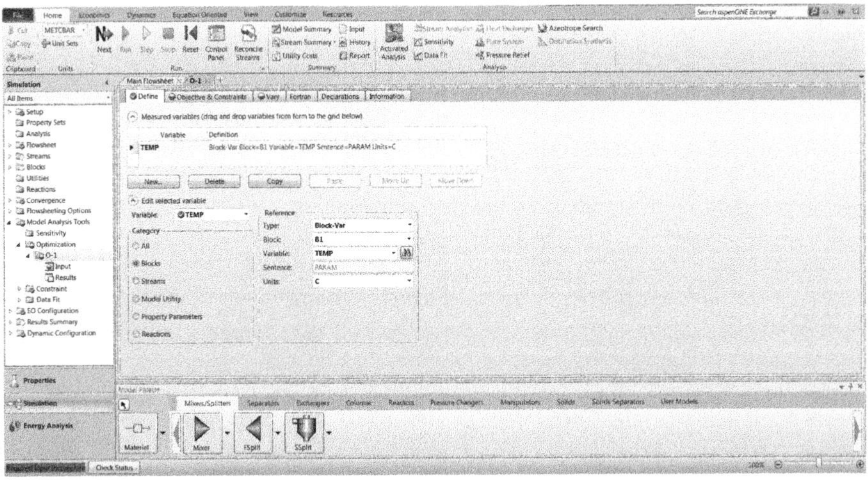

FIGURE 5.45
Define sheet for the temperature of the flash drum.

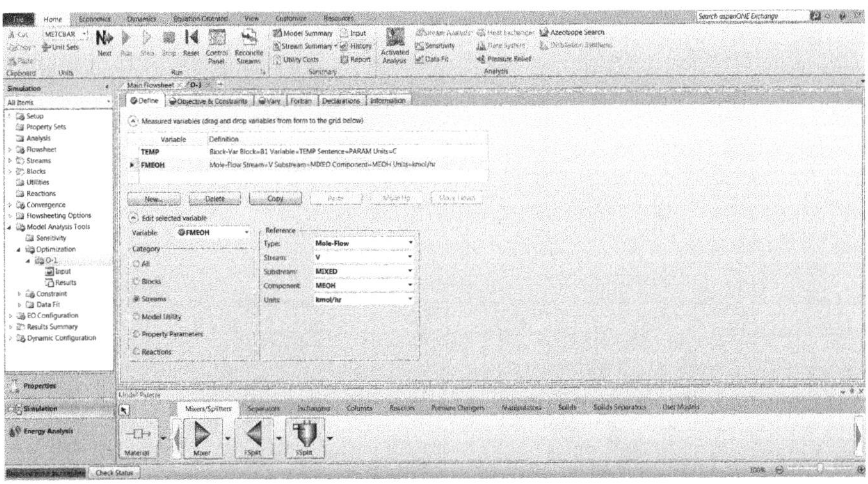

FIGURE 5.46
Define sheet for the mole flow rate of methanol in the vapor stream of the heat exchanger.

sheet. The variable FMEOH is selected as the objective function, which will be maximized. Below the space where the objective function is defined, there is an area to select the constraints of the problem. Nevertheless, that space is blank (Figure 5.48). This is because the constraints are defined in other subfolder, which is called *Constraint*, and is also located in the *Model Analysis Tools* folder. If the *Constraint* subfolder is opened, a window where a new constraint must be defined will be opened. When the *New* button is pushed, the ID of the constraint appears as C-1. This default name is

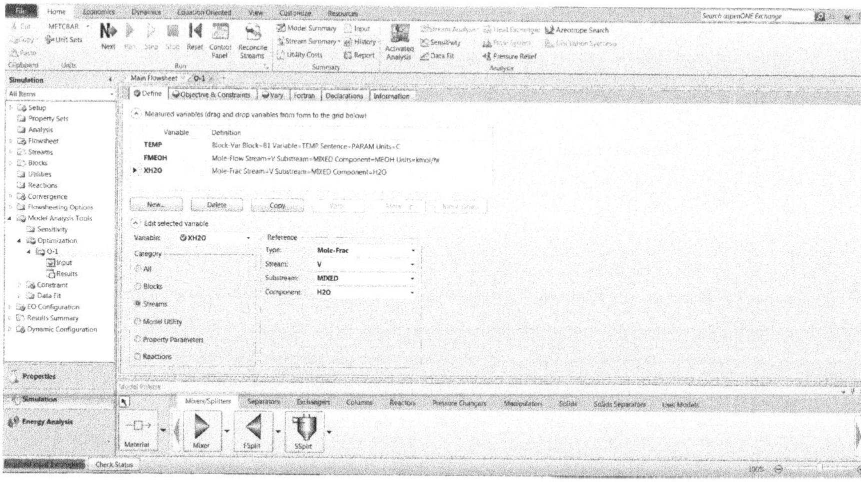

FIGURE 5.47
Define sheet for the mole fraction of water in the vapor stream of the heat exchanger.

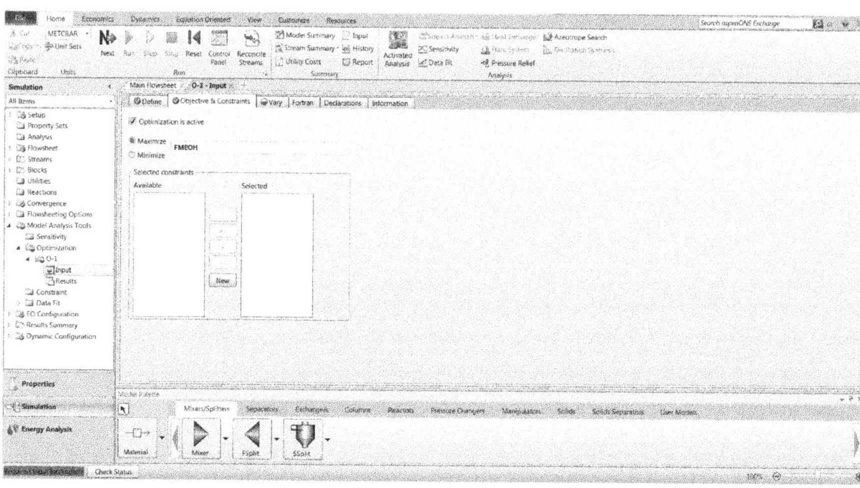

FIGURE 5.48
Objective function for the flash drum.

used for our constraint (Figure 5.49). Now, the variables involved in the constraint are established in the *Define* sheet. In this case, the constraint involves the purity of water, thus the measured variable is X_{H_2O}, whose characteristics were defined in the *Optimization* subfolder. Now, in the *Spec* sheet the constraint is identified. There are three lines on this sheet. In the first one, at the top, the name of the variable is written. In the middle line, at the left, the type of constraint is selected from a menu. The constraint to be defined in this case is *Less than or equal to*. In the same line, next to the

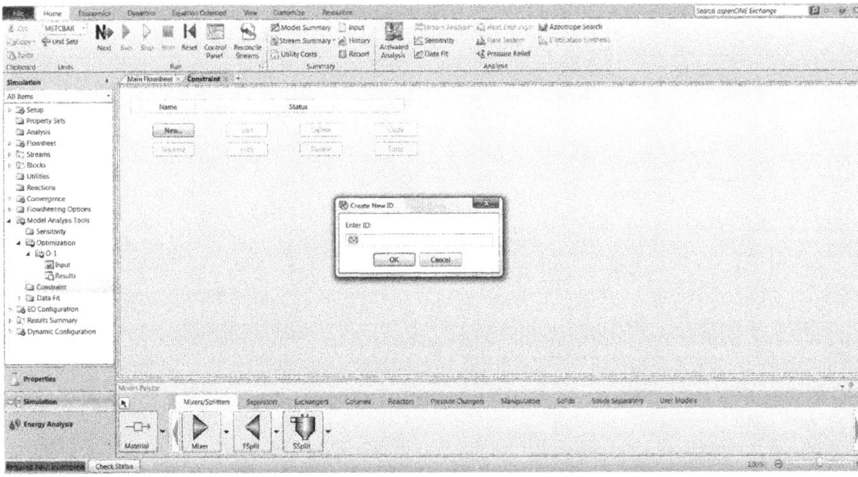

FIGURE 5.49
Creation of a new constraint for the flash drum.

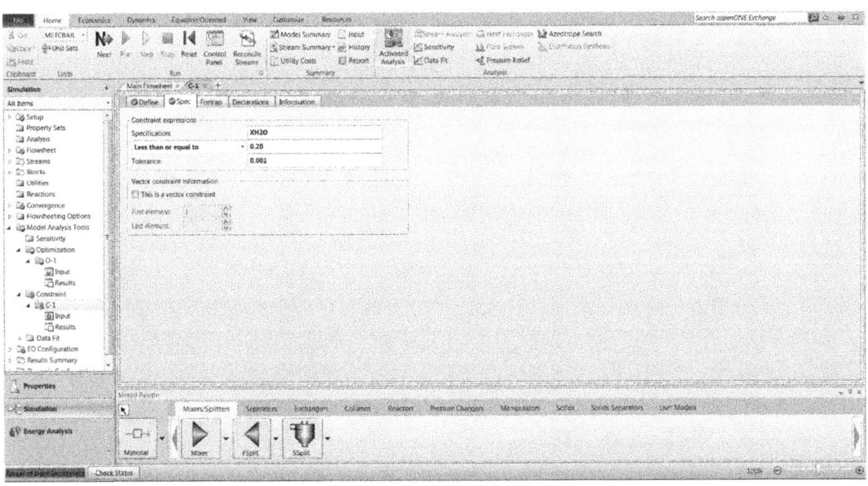

FIGURE 5.50
Specification of the constraint for the flash drum.

menu, the numeric value at the right of the constraint is defined. In this case, that value is 0.20. This information represents the constraint $y_{H_2O} \leq 0.20$. In the bottom line, the tolerance is defined. Here, we used a value of 0.001 for tolerance (Figure 5.50). With this, the constraint definition has been completed. Now, we must return to the *Objective & Constraints* sheet in the O-1 subfolder. In the *Selected constraints* box, it can be observed that the new constraint C-1 appears. It is selected and moved to the box in the right (Figure 5.51). Finally, the manipulated variable is defined in the *Vary*

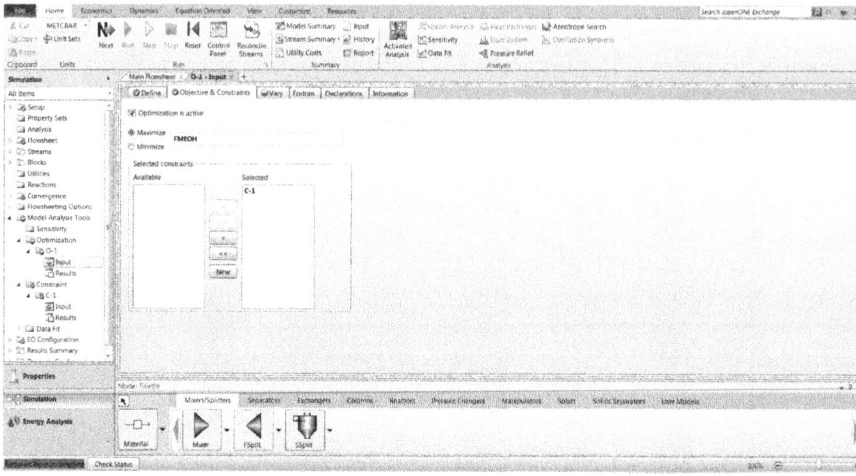

FIGURE 5.51
Activation of the constraint for the flash drum.

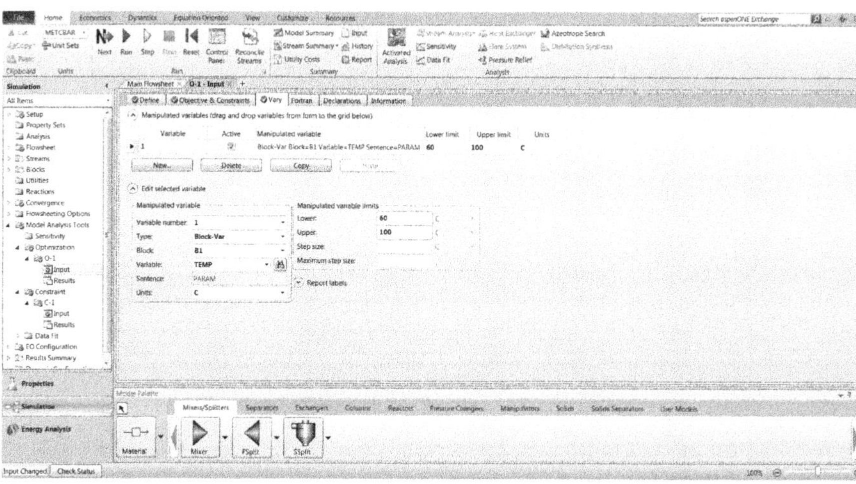

FIGURE 5.52
Vary sheet for the flash drum.

sheet. The *New* button is pushed to obtain a new manipulated variable. For the flash drum, that variable is the temperature, which is a block variable (*Block-Var*). Once more, the lower and upper limits of the variable are fixed as 60°C and 100°C, respectively (Figure 5.52). All the required information has been now loaded; thus, the *Next* button is pushed to run the simulation. Results for the optimization procedure are presented in Figure 5.53. The maximum flow rate of methanol in the vapor phase (22.3541 kmol/h)

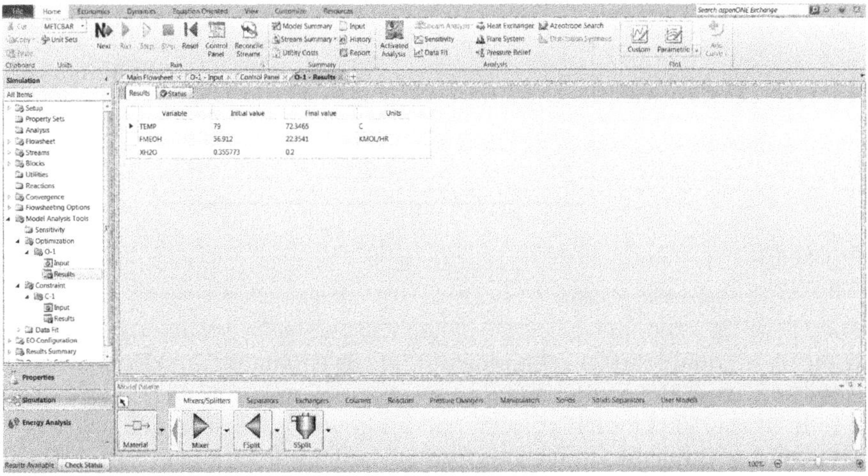

FIGURE 5.53
Results for the optimization of the flash drum.

occurs at the upper bound of the water composition when the flash temperature is 72.34°C. The result obtained through the SQP method is more accurate than that obtained by the sensitivity analysis because the results obtained through the sensitivity analysis depend on the number of points.

5.6 Optimization of a Tubular Reactor

5.6.1 Description of the Problem

In this case of study, the esterification of lactic acid with methanol takes place in a tubular reactor. The reaction occurs as follows:

$$\text{LACTAC} + \text{MEOH} \leftrightarrow \text{METLACT} + H_2O \tag{5.1}$$

where LACTAC represents the lactic acid; MEOH is the methanol; METLACT is the product of interest, methyl lactate; and H_2O represents water. The reaction is usually catalyzed by ion-exchange resins, but it has been reported that the reaction can be autocatalyzed using the acid for temperatures higher than approximately 360 K (Sanz et al., 2002). This approach is used in this example. The kinetic model reported by Sanz et al. (2002) is used to represent the reaction rate, but the effect of the reverse reaction is neglected. Thus, the reaction rate is expressed as follows:

$$r = k_1 \exp\left(\frac{-E_A}{RT}\right) a_{\text{LACTAC}}^2\, a_{\text{MEOH}} \tag{5.2}$$

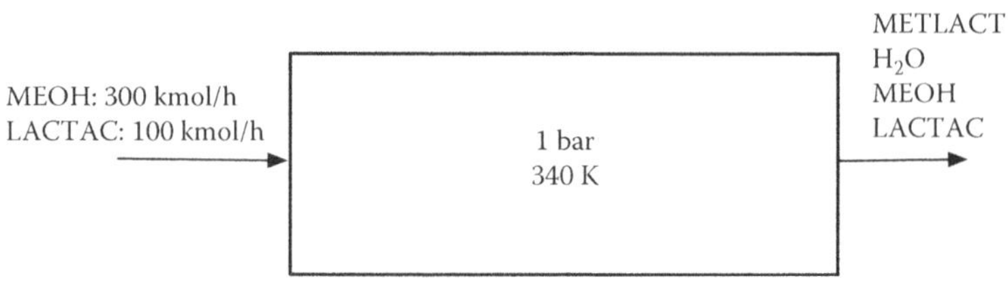

FIGURE 5.54
Tubular reactor.

where k_1 is equal to 6.024×10^8 mol·min^{-1}, E_A is equal to 56.45 kJ·mol^{-1}, and a_i is the activity of the component i.

In this case, 100 kmol/h of lactic acid is allowed to enter a tubular reactor. Methanol is also fed to the reactor, with a methanol/lactic acid mole ratio of 3:1. The reactor operates at 340 K and 1 bar. Esterification of lactic acid occurs inside the vessel; thus, the stream leaving the equipment contains methyl lactate, water, and unreacted methanol and lactic acid. The process is represented in Figure 5.54.

In this case, the objective is to maximize the following function:

$$Z = F_{\text{METLACT}} C_{\text{METLACT}} - F_{0,\text{LACTAC}} C_{\text{LACTAC}} - \frac{C_{\text{REACT}}}{n} \quad (5.3)$$

where F_{METLACT} is the mass flow rate of methyl lactate leaving the reactor (kg/year), C_{METLACT} is the unitary sales cost of methyl lactate, $F_{0,\text{LACTAC}}$ is the mass flow rate of lactic acid entering the reactor (kg/year), C_{LACTAC} is the unitary cost of lactic acid, C_{REACT} is the cost of the reactor, and n is the recovery time of the investment (years). In this problem, C_{METLACT} is assumed to be 12,700 USD/kg, whereas C_{LACTAC} is assumed to be 135 USD/kg. These costs have been estimated on the basis of those reported by Sigma-Aldrich (www.sigmaaldrich.com). The cost of the reactor is approximated as follows:

$$C_{\text{REACT}} = 4{,}000L \quad (5.4)$$

where L is the length of the reactor, in meters. Finally, n has been assumed as 5 years.

5.6.2 Initial Simulation

In this section, the initial simulation of the reactor is described. Four components (lactic acid, methanol, methyl lactate, and water) are defined in the *Components—Specifications* sheet (Figure 5.55). The UNIFAC model is selected in the *Methods—Specifications* sheet (Figure 5.56) to represent the vapor–liquid equilibrium. This information is sufficient to run the properties test. Because the test is positive, we now move toward the simulation section, where the flowsheet is defined. An *RPlug* module is selected from the *Reactors* folder, representing the tubular reactor, and inserted in the flowsheet. The feed and product streams are also inserted (Figure 5.57). The conditions of the feed stream are now defined. Feed temperature and pressure are set equal to those of the reaction. As aforementioned, feed stream has 100 kmol/h of lactic acid and 300 kmol/h of methanol. The *Input* sheet for the feed stream can be viewed in Figure 5.58. As the next step, the characteristics of the reactor are loaded, in the module *Setup* sheet. First, the

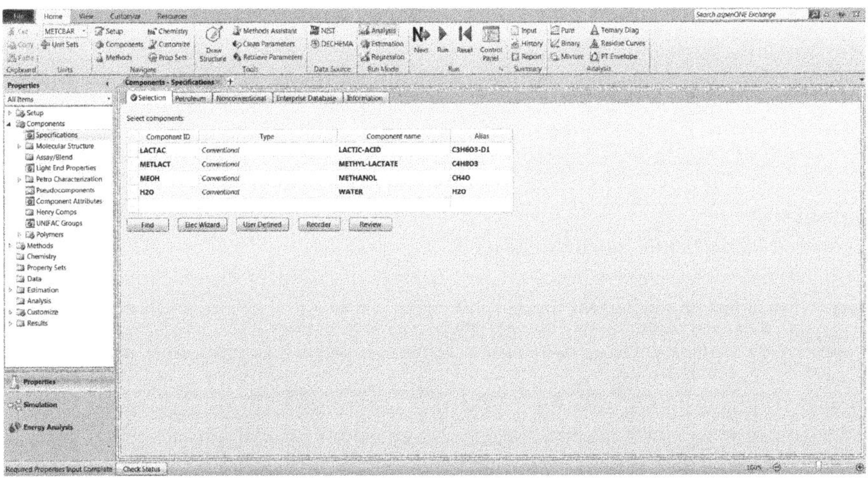

FIGURE 5.55
Selection of the components for the tubular reactor simulation.

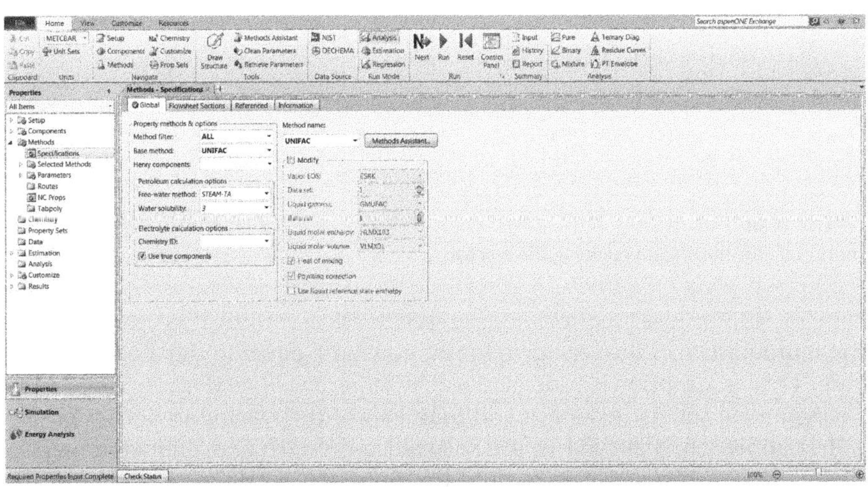

FIGURE 5.56
Selection of thermodynamic method for the tubular reactor simulation.

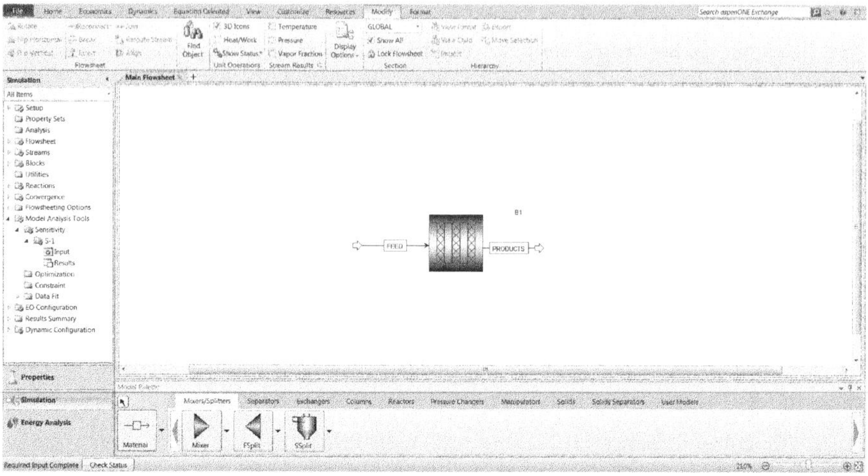

FIGURE 5.57
Flowsheet for the tubular reactor simulation.

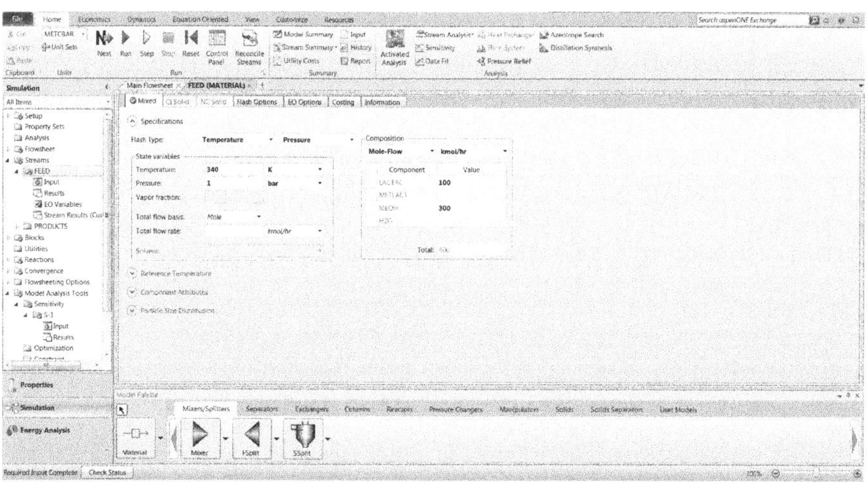

FIGURE 5.58
Input data for the stream entering the reactor.

type of reactor is set as *Reactor with specified temperature*. It is assumed that the temperature of the reactor remains constant, equal to the feed temperature. This assumption is indicated as *Operating condition* (Figure 5.59). In the *Configuration* sheet, the length and diameter of the reactor are indicated. As a first approach, 5 and 0.1 m are selected for length and diameter, respectively. The reaction is assumed to occur in the liquid phase. Hence, the phase is indicated as the only valid phase in the reactor (Figure 5.60). The next sheet to be completed is the *Reactions* sheet. Nevertheless, no available

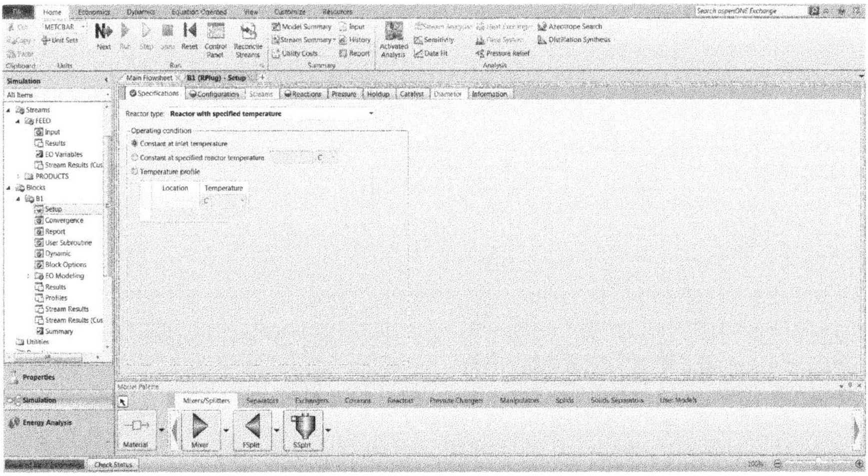

FIGURE 5.59
Specifications sheet for the reactor.

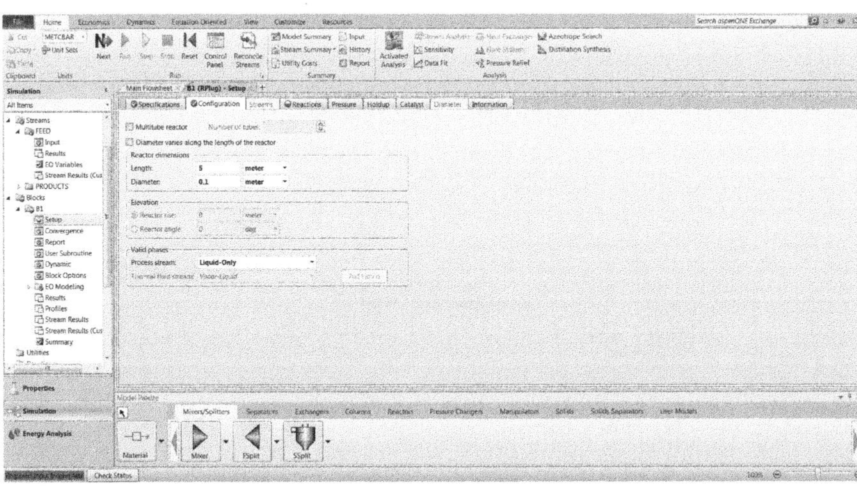

FIGURE 5.60
Configuration sheet for the reactor.

reactions are to be selected (Figure 5.61) because they are defined in the *Reactions* folder, which is located below the *Blocks* folder, at the menu on the left side of the screen. Once the *Reactions* folder is opened, a new reaction pack is defined, and an ID is assigned to it. In this case, the default ID *R-1* is used. Because the kinetic model is provided in the form of a power law, the reaction type is selected as *POWERLAW*, as shown in Figure 5.62. Now, the information about the reaction is loaded. In the *Stoichiometry* sheet, the reaction type is selected as *Kinetic*, the coefficients of the reactants (lactic

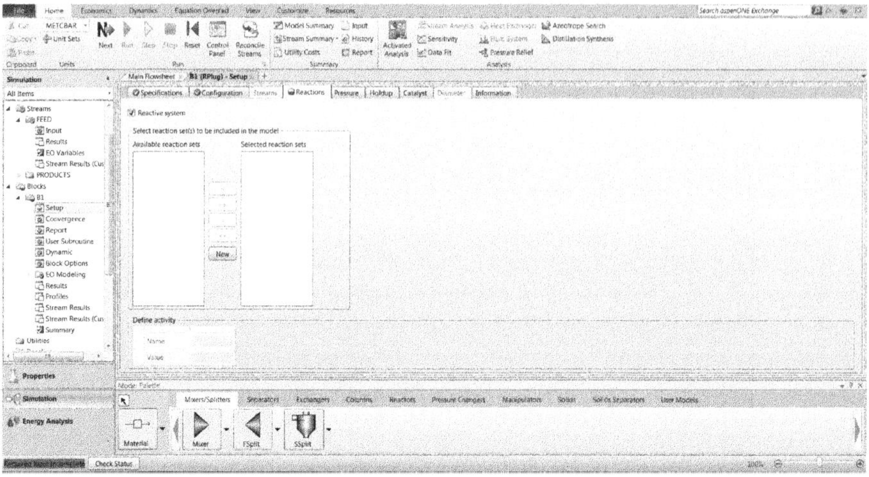

FIGURE 5.61
Reactions sheet for the reactor.

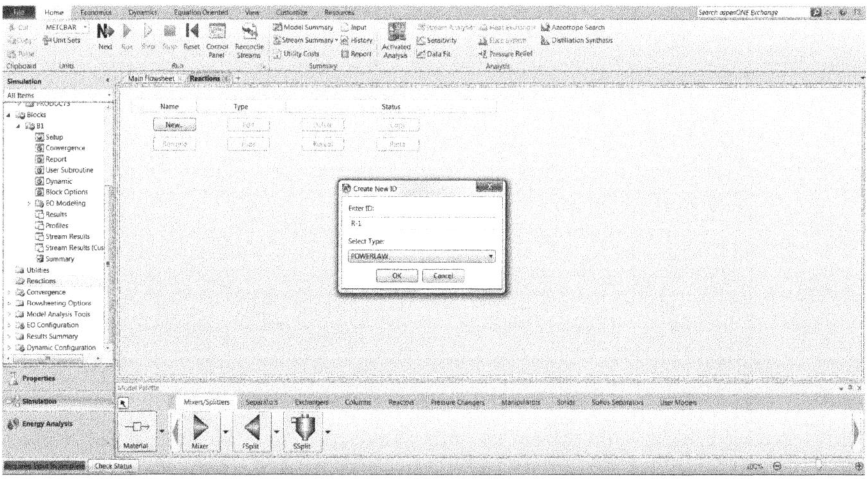

FIGURE 5.62
Creation of a new power law reaction.

acid and methanol) are set as –1, and the coefficients of the products (methyl lactate and water) are set as 1. The exponent is defined as 2 for lactic acid and 1 for methanol because it is the power for each component activity in the kinetic expression. The products are not involved in the kinetic model for the straight reaction. Thus, the field is left blank, which implies an exponent of zero (Figure 5.63). Now, the kinetic data of the reaction are loaded in the *Kinetic* sheet. The values for the pre-exponential factor *k* and the energy of activation *E* are known, and they are directly loaded in the sheet. As

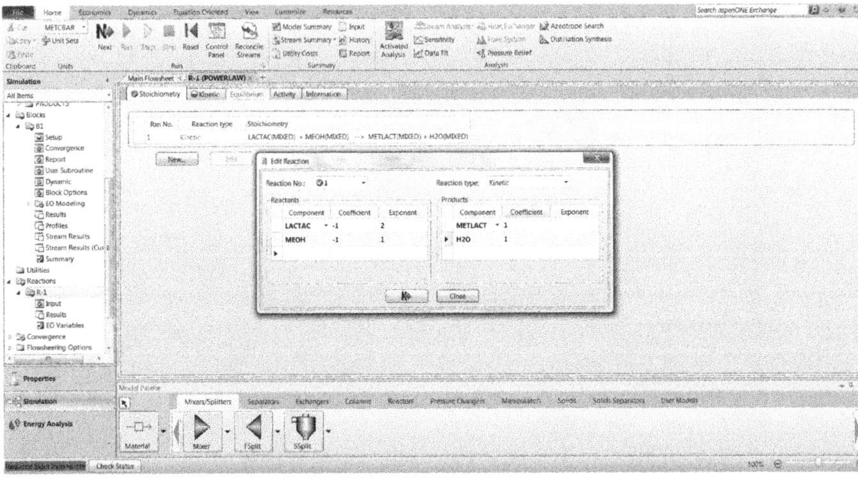

FIGURE 5.63
Definition of the stoichiometry of the reaction.

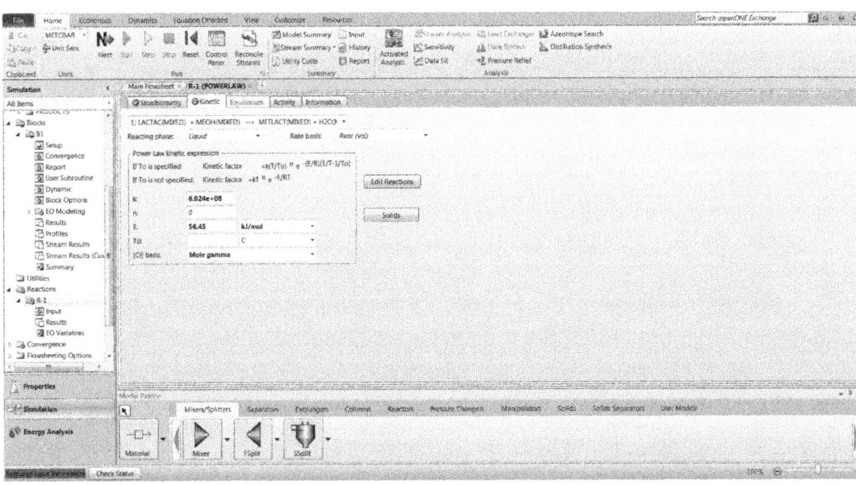

FIGURE 5.64
Definition of the kinetic parameters of the reaction.

the reference temperature T_0 is not specified, the exponent n is set as 0 to allow the dependence of temperature to be modeled using the Arrhenius equation. Finally, the basis for concentration is selected as *Mole gamma* to allow the kinetic calculations to be based on the activities of the components. This sheet can be observed in Figure 5.64. Once the reaction has been defined, we can go back to the *Reaction* sheet in the *Setup* menu of the block. The reaction R-1 is now selected and sent to the *Selected reaction sets* box (Figure 5.65). After loading all the required information, the *Run* button is

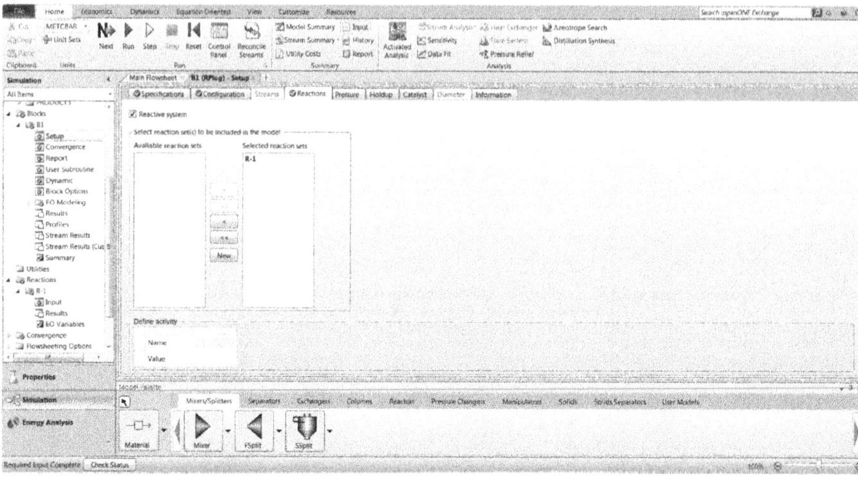

FIGURE 5.65
Activation of the reaction for the reactor.

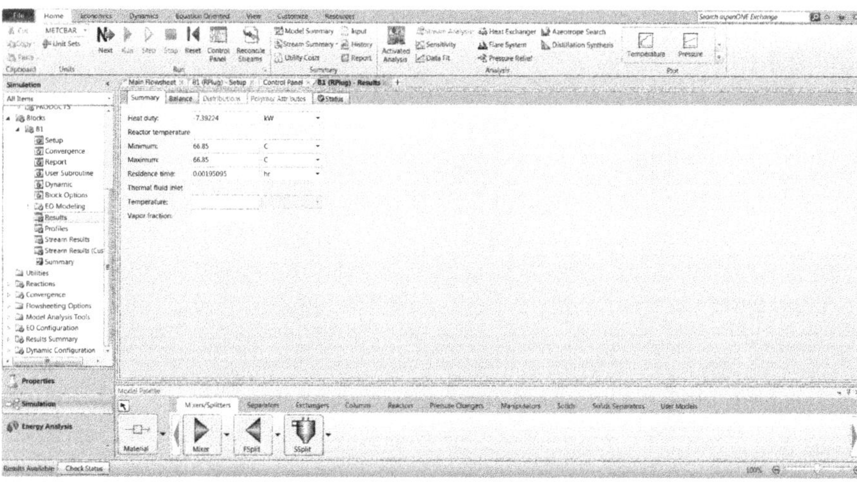

FIGURE 5.66
Results summary for the initial simulation of the reactor.

pushed to start the simulation. Figure 5.66 shows the results for the reactor. The reaction releases energy, and the residence time becomes considerably low. Nevertheless, at this step, it is of more interest to determine how much product has been obtained. This can be observed in the *Stream Results* sheet (Figure 5.67), where a considerably low conversion to methyl lactate can be noticed, which is approximately 3.3 mol%. Thus, the first assumption of the design characteristics of the reactor is not effective in terms of the conversion to methyl lactate. Better design alternatives are expected after the sensitivity analysis.

5.6.3 Optimization through Sensitivity Analysis

In the case of the reactor, the objective function depends on the flow rate of methyl lactate at the exit of the reactor and on the length of the reactor. Feed flow rate of lactic acid is known (100 kmol/h or 9008 kg/h); thus, the term $F_{0,LACTAC}C_{LACTAC}$ can be easily computed as 1.045×10^{10} USD/year, assuming 8600 operating hours per year. This value can be substituted in the objective function.

Now, the sensitivity analysis is performed. As for the previous examples, a new ID is created in the *Sensitivity* subfolder, naming the analysis folder as S-1. The length and diameter of the reactor are selected as manipulated variables. This information is loaded in the *Vary* sheet. Now, the information required in the *Vary* sheet must be completed. Both variables are classified as block variables (*Block-Var*) defined as *LENGTH* and *DIAM*, respectively. The limits for the reactor length are set as 1 and 200 m, whereas the diameter ranges between 0.1 and 2.1 m. The information for each manipulated variable is shown in Figures 5.68 and 5.69, respectively. Moreover, the measured variable is defined. For the reactor, the mass flow rate of methyl lactate (FMETLAC) in the outlet stream is the measured variable. The variable can be observed in the *Streams* category, with a type *Mass-Flow*, as shown in Figure 5.70. Finally, in the *Tabulate* sheet, the *Fill Variables* button is pushed to indicate the variables to be reported (Figure 5.71). Then, the simulation occurs by pushing the *Next* button. The *Results* sheet is shown in Figure 5.72. Here, the changes on the flow rate of the main product

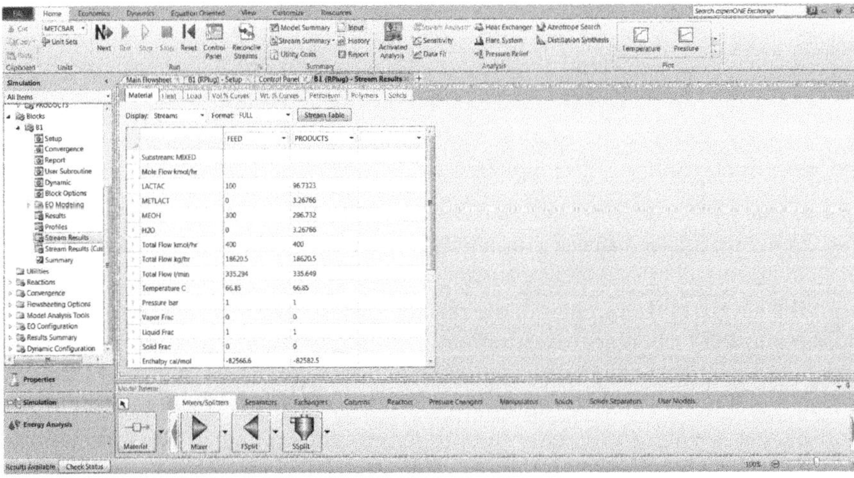

FIGURE 5.67
Streams results for the initial simulation of the reactor.

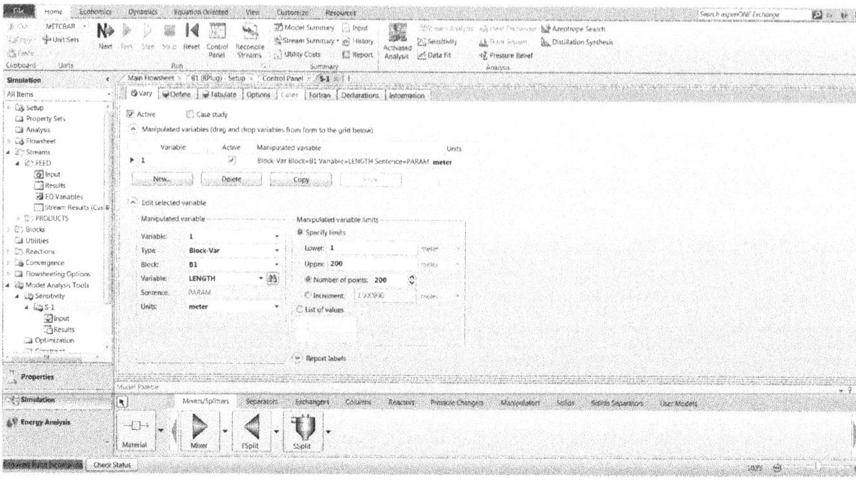

FIGURE 5.68
Definition of the reactor length as manipulated variable for the sensitivity analysis.

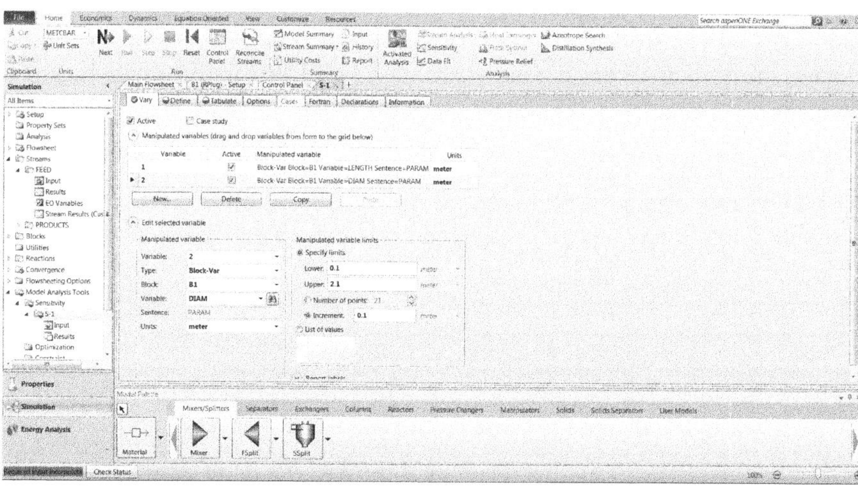

FIGURE 5.69
Definition of the reactor diameter as manipulated variable for the sensitivity analysis.

with the length and diameter of the reactor are observed. Nevertheless, it is not possible to load the objective function for this case of study; thus, its value is computed using a second software, such as Microsoft Excel. A plot with some selected solutions is shown in Figure 5.73. It is clear that the objective function increases with an increase in both length and diameter. Nevertheless, as the diameter approaches 2 m, the variation in the objective function becomes smaller. Moreover, when the length of the reactor varies from 20 to 30 m, the objective function results are similar, particularly, for diameters close to 2 m.

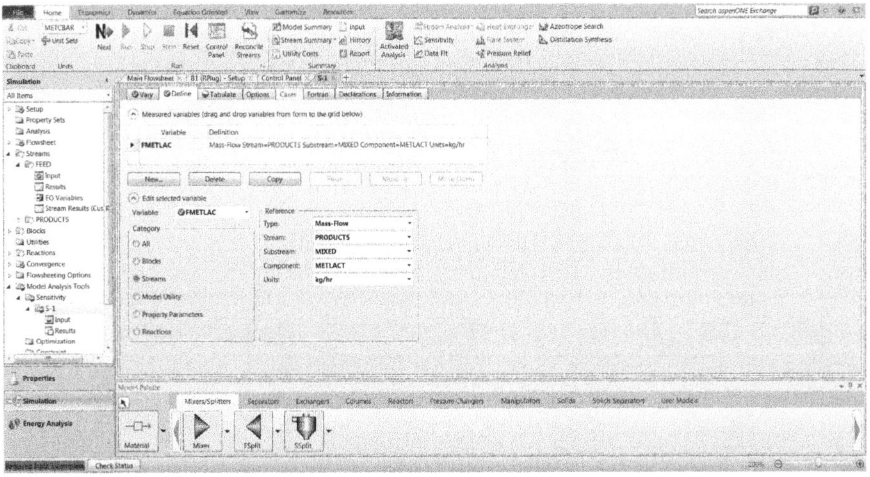

FIGURE 5.70
Definition of the methyl lactate flow rate as measured variable.

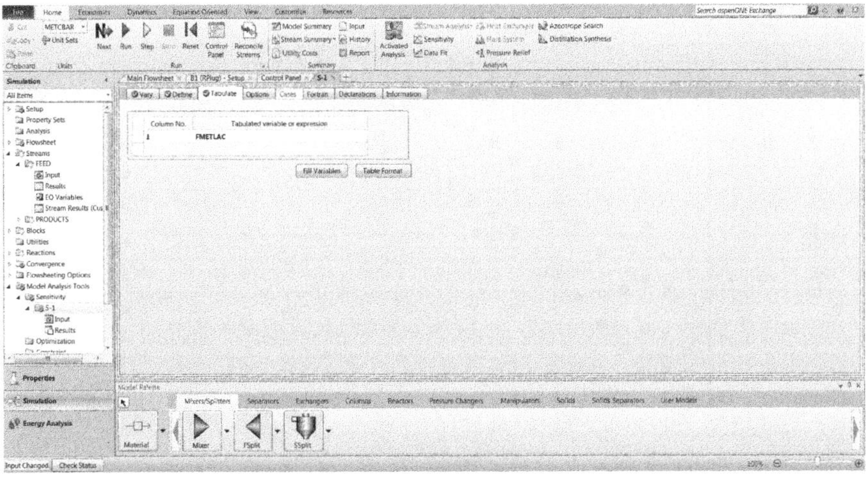

FIGURE 5.71
Tabulate sheet for the sensitivity analysis of the reactor.

Thus, in this stage, a design with 2 m diameter and 20 m length can be selected as the best one, for which $Z = 1.0911 \times 10^{12}$ USD/year.

5.6.4 Optimization through Sequential Quadratic Programming

The optimization of the reactor using the SQP method is presented in this section. As for the previous cases, a new optimization routine is created in the *Optimization* subfolder, which is identified with the ID O-1. Now, it is necessary to define the measured variables. The length of the reactor is first defined, which

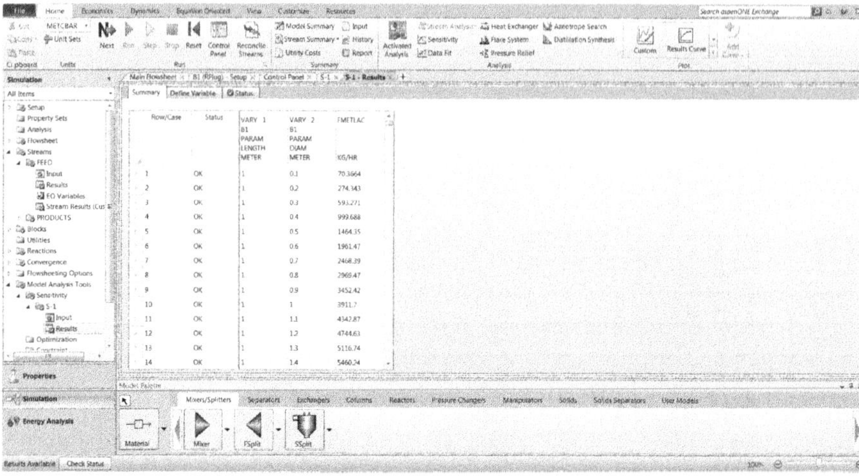

FIGURE 5.72
Results summary for the sensitivity analysis of the reactor.

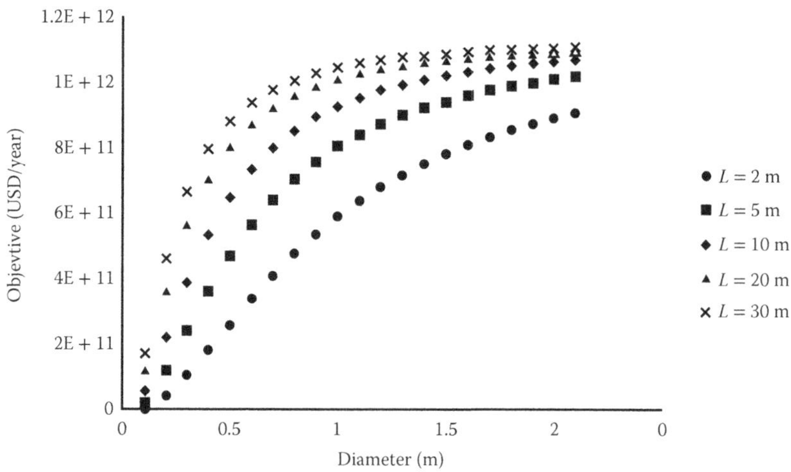

FIGURE 5.73
Variation of the objective function with the reactor length and diameter.

belongs to the *Blocks* category, with the type *Block-Var*, as shown in Figure 5.74. The diameter of the reactor is identified with the name *DIAMETER*, also belonging to the category *Blocks*, with a type *Block-Var*, defined by the software with the variable tag *DIAM* (Figure 5.75). The mass flow rate of methyl lactate in the outlet stream is named as *FMETLAC*, being in the *Streams* category, with a type *Mass-Flow* for the stream defined by the user as *PRODUCTS* (Figure 5.76). The objective function is now defined in the *Objective & Constraints* sheet. Nevertheless, in this case, the objective function is not a variable provided by

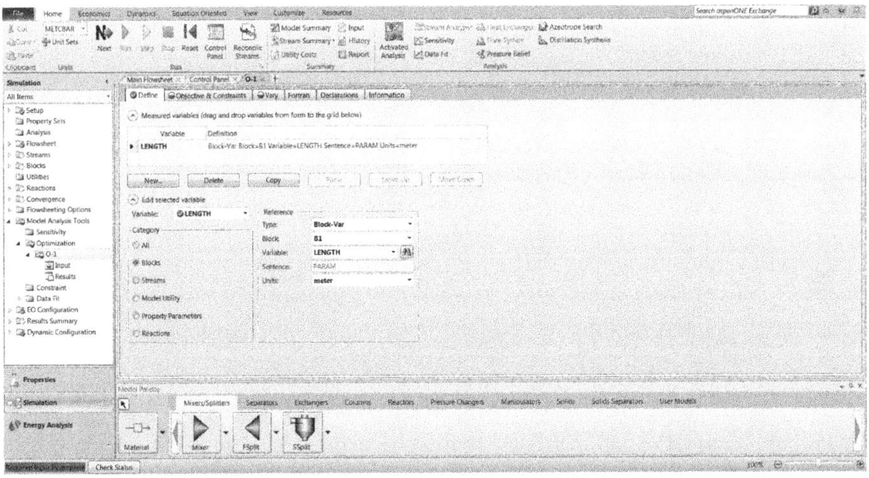

FIGURE 5.74
Define sheet for the reactor length.

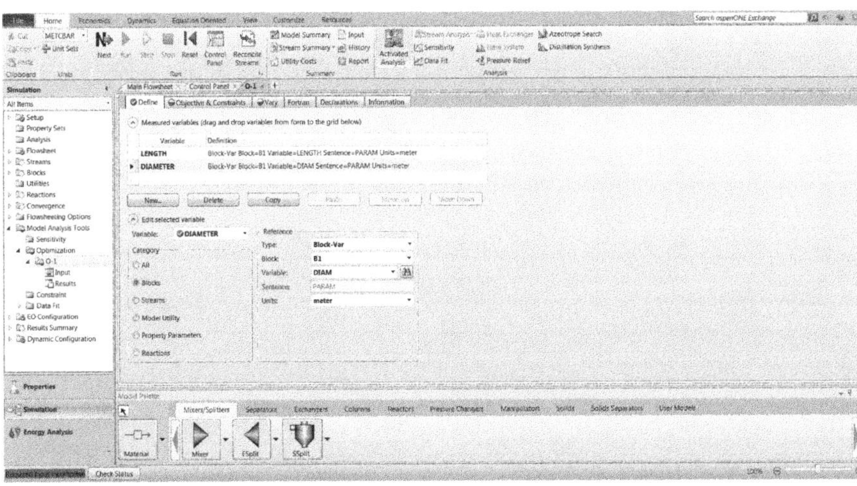

FIGURE 5.75
Define sheet for the reactor diameter.

the software but a combination of variables. Thus, in the *Objective & Constraints* sheet, only the name of the variable representing the objective function (i.e., Z) and the type of optimization to be performed (i.e., maximization) is indicated, as shown in Figure 5.77. The objective function is defined as a routine in the *FORTRAN* sheet, where all the numeric values of the parameters are also defined (Figure 5.78). Now, the manipulated variables are defined in the *Vary* sheet. The first manipulated variable is the length of the reactor, whose lower and upper limits are defined as 1 and 200 m, respectively (Figure 5.79). The

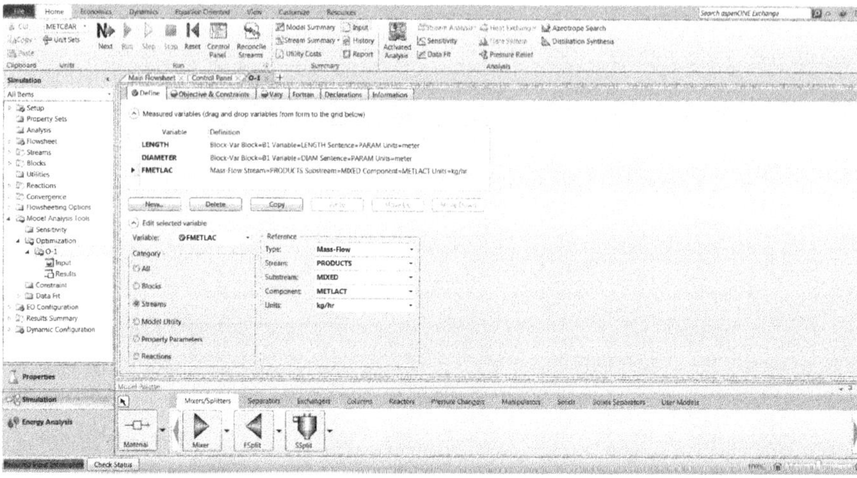

FIGURE 5.76
Define sheet for the mass flow rate of methyl lactate at the products stream.

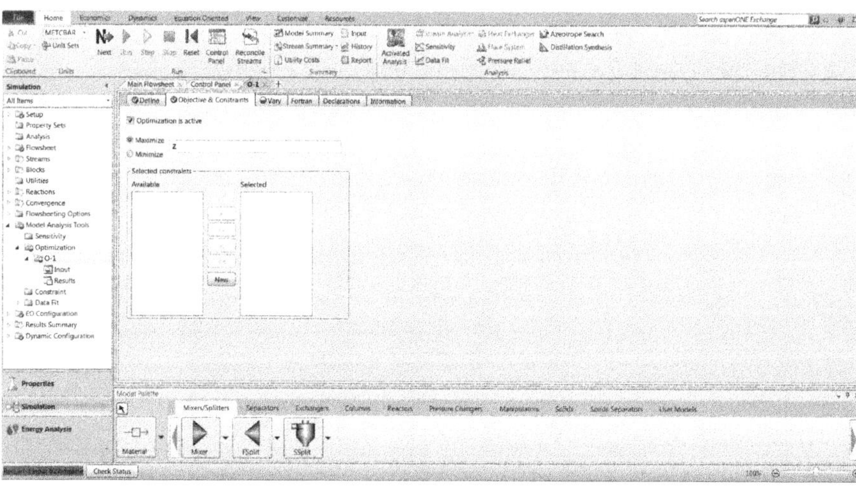

FIGURE 5.77
Objective function for the reactor.

second manipulated variable is the diameter, for which the limits are defined as 0.1 and 2.1 m, respectively (Figure 5.80). Now, all the required information is loaded and the *Next* button is pushed to run the simulation. Results for the optimization procedure are shown in Figure 5.81. In the *Final value* column, a reported solution of L of 200 m and D of 2.1 m, with a mass flow rate of methyl lactate of 10,378.1 kg/h, which is equivalent to 99.69 kmol/h, is defined. Thus, a conversion of 99.69% from lactic acid to methyl lactate has been obtained. With the reported data, it can be computed that the maximum value for Z is

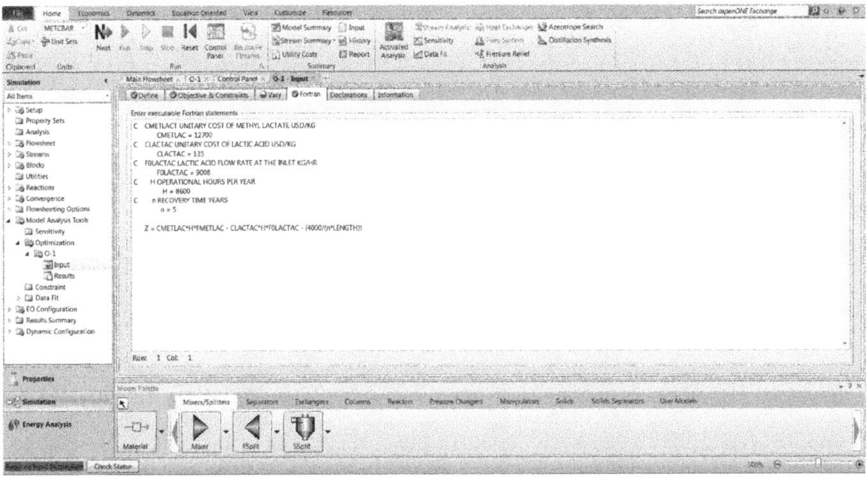

FIGURE 5.78
FORTRAN routine for the calculation of the objective function.

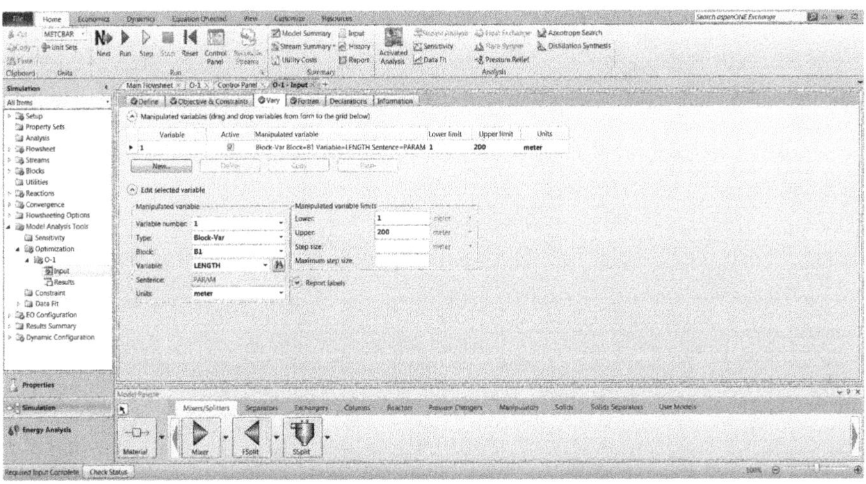

FIGURE 5.79
Definition of the reactor length as manipulated variable.

1.123×10^{12} USD/year. This solution differs from that obtained through sensitivity analysis by only 2.8%, but requires a considerably larger reactor. This is because the objective function does not show a strict maximum, and Z increases as the length and diameter of the reactor are increased. Thus, additional criteria are required to select the best solution. Those criteria could be added as constraints for the optimization problem.

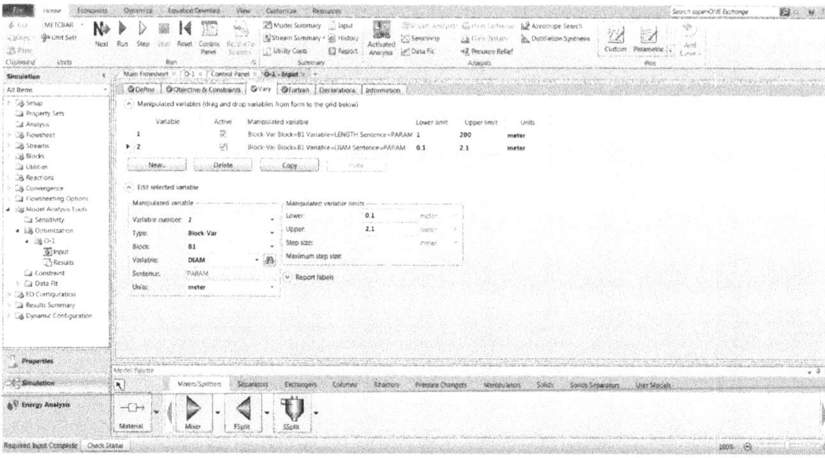

FIGURE 5.80
Definition of the reactor diameter as manipulated variable.

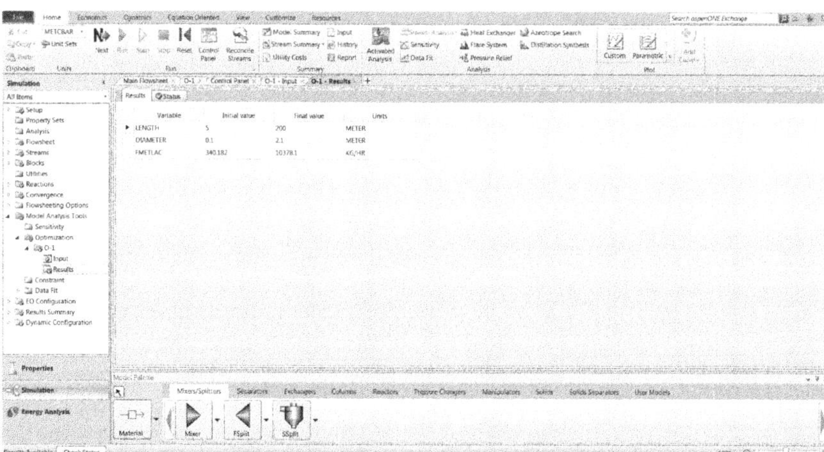

FIGURE 5.81
Results for the optimization of the reactor.

References

P.E. Gill, W. Murray, M.A. Saunders, 2005, SNOPT: An SQP algorithm for large-scale constrained optimization, *SIAM Rev.*, 47(1), 99–131.

M.T. Sanz, R. Murga, S. Beltrán, J.L. Cabezas, 2002, Autocatalyzed and ion-exchange-resin-catalyzed esterification kinetics of lactic acid with methanol, *Ind. Eng. Chem. Res.*, 41(3), 512–517.

6

*Optimization using Aspen Plus® and Stochastic Toolbox**

6.1 Introduction

This chapter discusses a complete methodology that allows optimizing any process modeled in Aspen Plus®. The essence of the method is based on the capacity of Aspen Plus, MATLAB, and Microsoft Excel to be linked, exploding the capacities of each software. Therefore, it is possible to take information (vectors) from one of those programs to other.

Nowadays, a clear tendency is to evaluate several objectives, such as economic and environmental, simultaneously. Several studies are the proof of this trend (Li et al., 2014; Wang et al., 2007). Under this scenario, both mono-objective and multi-objective optimizations are analyzed, considering a couple of distillation columns as example. However, this optimization process is not restricted for only distillation columns. It could be performed for any type of module within Aspen Plus.

6.2 Software for Stochastic Optimization

In Chapter 2, procedures to solve several engineering problems from a deterministic point of view have been described. In this chapter, a stochastic approach is presented. This type of approach helps us to solve several potentially nonconvex and highly nonlinear problems (Segovia-Hernández et al., 2015).

A multifaceted program, very commonly used today is MATLAB (stands for Matrix laboratory). This is currently one of the most widely used software in several knowledge areas; engineering is not an exception. It allows manipulation of any type of matrix, plotting of functions and data, and also creation and implementation of algorithms. Further, it allows creating user interfaces and interfacing with external programs where a code could be written with other type of languages, such as C, C++, and FORTRAN.

* Eduardo Sánchez-Ramírez and Juan José Quiroz-Ramírez also contributed to the work on this chapter.

Inside MATLAB, it is possible to find several toolboxes and packages. Regarding toolboxes, the optimization toolbox provides considerable information about solvers, previously written to find parameters that minimize or maximize objectives with or without constraints. The optimization toolbox offers some stochastic optimization algorithms, such as simulated annealing and genetic algorithms in both mono-objective and multi-objective optimization. It is quite simple to find the optimization toolbox. Figure 6.1 shows the exact location of this toolbox. Notice that all figures where MATLAB is used were obtained with MATLAB 2013b (8.2.0.701).

FIGURE 6.1
Optimization toolbox inside MATLAB.

6.3 Linking Aspen Plus with the Stochastic Optimization Software

To link MATLAB as the main program to optimize any model in Aspen Plus is indeed an easy task; however, it is important to do this in an orderly manner. Before starting the entire explanation, it is essential to note that MATLAB is the main software in the entire optimization process, which indicates that this program will be open during the entire process.

Initially, the stochastic algorithm, which is presented in the optimization toolbox in MATLAB, proposes the data to be evaluated in our model. To perform this task, MATLAB uses Microsoft Excel to transport data as vectors from MATLAB to Aspen Plus, and Aspen Plus evaluates the vectors and produces the classical results in Aspen Plus, such as reboiler heat duty, condenser heat duty, molar flows, and molar fractions. Those data returned as vectors to Microsoft Excel can be used to calculate any objective functions such as the total annual cost and CO_2 emissions. Finally, the objective or objectives are then returned to MATLAB, which evaluates those objective functions and proposes new vectors to be evaluated again, until a stop criterion is achieved. Figure 6.2 illustrates a simple flowsheet of the process described previously.

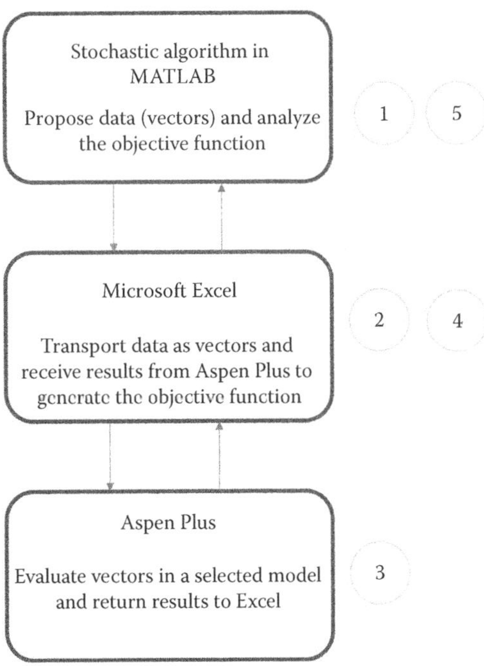

FIGURE 6.2
Simplified flowchart of the optimization process.

However, in this chapter, the functioning of this interface is explained in general terms. Nevertheless, each step contains considerable details, which are not the main target of this chapter. For example, the generation of data from the stochastic algorithm totally depends on the nature of the algorithm; in other words, simulated annealing generates vectors in a considerably different manner than genetic algorithms or differential evolution.

6.3.1 Creating a Function to be Optimized with MATLAB

As mentioned earlier, during the optimization process, MATLAB is the main program because the algorithm is already written within MATLAB. Therefore, the first step before optimization is to create a function. The purpose of this function is "to let it know" to MATLAB where all the files are located; in other words, MATLAB will know the exact route of Excel File and also where it will write the produced vectors. The entire code of this function is as follows:

```
function f=funobj(X)
new=X; %variables vector
%Normalization of values
new(1,1)=new(1,1)*100;
new(1,2)=new(1,2)*30;
new(1,3)=new(1,3)*100;
new(1,4)=new(1,4)*30;
%round values
new(1,1)=round(new(1,1));
new(1,3)=round(new(1,3));
xlswrite('C:\Users\PC\Documents\Temperature.xls',
new,'Temperature','G7:J7')
h = actxserver('Excel.Application');
wkbk = h.Workbooks;
file = wkbk.Open('C:\Users\PC\Documents\Temperature.xls');
h.visible=1;
Application = wkbk.Application;
ActiveSheet=h.Activesheet;
invoke(Application,'Run','annealing');
h.WorkBooks. Close
h.Quit;
h.delete;
f=xlsread('C:\Users\PC\Documents\Temperature.xls',new,'Tempera
   ture','Temperature','AH7');
```

There are numerous commands inside the codes of this function. We highlight some of them. We named our vector "new," and it is shaped by four elements (1,1) to (1,4). These four elements are the variables in the model to be optimized, for example, total stages, reflux ratio, distillate to feed ratio, etc.

However, some variables are discrete (stages, feed stage, etc.) and hence we defined later new (1,1) and new (1,3) as integer numbers by adding the command "round." In the command "xlswrite," the entire route of the Excel file must be written. Moreover, it is necessary to specify the cells where MATLAB will write, for example, the cells are G7 to J7. On the other hand, in the command "file = wkbk.Open" the entire route must be written. The line "h.visible=1" could be changed between 0 and 1; with a value of 0 Microsoft Excel will function in the second plane whereas with a value of 1 the display of Microsoft Excel will appear continuously during optimization. In the command "invoke(Application,'Run','annealing')," "'annealing' indicates that within Microsoft Excel a subroutine with this name is written. This subroutine will be analyzed further. Finally, the last line "f=xlsread('C:\Users\PC\Documents\Temperature.xls', new,'Temperature','Temperature', 'AH7')" is actually the route where the objective function is located; in this case, the cell AH7 contains its value. It is necessary to remark that the value of this cell always remains minimized. Same happens in both mono-objective and multi-objective optimization; thus, if the value need to be maximized, it is necessary to minimize the negative value of the objective function.

6.3.2 Creating a Subroutine in Microsoft Excel

The second step is to create a subroutine inside Microsoft Excel using Visual Basic. It is essential to write a code that helps us to lead the data generated in the stochastic algorithm in MATLAB to Aspen Plus. Therefore, it is important to write the route of the Excel file correctly. Figure 6.3 shows the location of program in Microsoft Excel. Moreover, all Excel screen captures were taken from Microsoft Excel 2013.

Once we start working in this platform, it is essential to write the subroutine. A simple example of this type of code is as follows (this subroutine has been named annealing):

```
Sub annealing()
Set Aspen = GetObject("C:\Users\PC\Documents \File1-bkp")
Aspen.Visible = False
'SEPD
'Stages SEPD
Range("Temperature!G7").Select
Aspen.Tree.Data.Blocks.SEPD.Input.NSTAGE.Value = ActiveCell.
   Value

'Reflux ratio CI
Range("Temperature!H7").Select
Aspen.Tree.FindNode("\Data\Blocks\SEPD\Input\BASIS_L1").Value =
ActiveCell.Value
'Aspen.Tree.Data.Blocks.CI.Input.BASIS_RR.Value = ActiveCell.
   Value
```

```
'***************************************************************
************************
'SEPE
'Number of stages SEPE
Range("Temperature!I7").Select
Aspen.Tree.Data.Blocks.SEPE.Input.NSTAGE.Value = ActiveCell.
  Value
NCI = Range("Temperature!I7").Value
'Reflux ratio
Range("Temperature!J7").Select
Aspen.Tree.FindNode("\Data\Blocks\SEPE\Input\BASIS_L1").Value =
   ActiveCell.Value
'Aspen.Tree.Data.Blocks.CI.Input.BASIS_RR.Value = ActiveCell.
   Value

'Run aspen
Application.DisplayAlerts = False
Aspen.Engine.Run
'_______________________________________________________________
____________________________________________
'Output data
'Molar Flow B in stream WATERB
Range("Temperature!K7").Select
ActiveCell.Value = Aspen.Tree.Data.Streams.WATERB.Output.
  MOLEFLOW.MIXED.BUTANOL.Value

'Mass Flow B in stream BUTOL
Range("Temperature!L7").Select
ActiveCell.Value = Aspen.Tree.Data.Streams.BUTOL.Output.
  MASSFRAC.MIXED.BUTANOL.Value

'Molar Flow B in stream BUTOL
Range("Temperature!X7").Select
ActiveCell.Value = Aspen.Tree.Data.Streams.BUTOL.Output.
  MASSFLOW.MIXED.BUTANOL.Value
'***************************************************************
****************************
'Run status
'Error
Range("Temperature! O7").Select
ActiveCell.Value = Aspen.Tree.FindNode("\Data\Results Summary\
  Run-Status\Output\PER_ERROR").Value
ER = Range("Temperature!O7").Value

If ER = 0 Then
'Reboiler SEPD
Range("Temperature!S7").Select
ActiveCell.Value = Aspen.Tree.Data.Blocks.SEPD.Output.REB_
  DUTY.Value
'Reboiler SEPE
```

```
Range("Temperature!T7").Select
ActiveCell.Value = Aspen.Tree.Data.Blocks.SEPE.Output.REB_
  DUTY.Value
End If
'*************************************************************
****************************
'Objective function
fn = Range("Temperature!AH7").Value
If IsRunning = True Then
    IsRunning = False
End If
      Aspen.Close
 If ActiveSheet.Range("C1").Value >= 0 Then
    Range("A7:AV7").Copy
    Sheets("Results").Select
    Variable = Range("A1").Value
Range("B" & Variable).PasteSpecial
    Application.CutCopyMode = False
Range("A1").End(xlDown).Offset(1, 0).Select
ActiveCell.FormulaR1C1 = "=R[-1]C+1"
Range("A1").Select
ActiveCell.FormulaR1C1 = "=R[3]C+2"
Range("A1") = Range("A1").End(xlDown).Offset(0, 0).Row
Sheets("Temperature").Select
    End If

If ER = 0 Then
TempFileName = Range("Resultados!A1").Value
Aspen.SaveAs Filename:=TempFileName & ".bkp"
End If
ActiveWorkbook.Save
End Sub
```

In this code, several sections are highlighted. The code followed by " ' " refers to comments, the code starting in line 2 (Set Aspen) and finishing in line 26 (before the Run Aspen command) refers the input data led to Aspen Plus, and finally the code in italics refers to output data from Aspen Plus to Excel. As a brief description, in the line "Set Aspen = GetObject("C:\Users\PC\Documents\File1-bkp")," the entire route of the Aspen file, in this case File1, should be specified. Furthermore, the commands for the input data to Aspen Plus are quite important; for example, the cell value considered in Aspen Plus should be specified in the first line:

```
Range("Temperature!G7").Select;
Aspen.Tree.Data.Blocks.SEPD.Input.NSTAGE.Value = ActiveCell.Value
```

A value previously proposed by MATLAB appears in the cell G7 in the sheet "Temperature." Therefore, with this command, this value is guided to the Aspen route "Aspen.Tree.Data.Blocks.SEPD.Input.NSTAGE," which corresponds to the number of stages. Clearly, this route refers to the block SEPD. Obviously, the

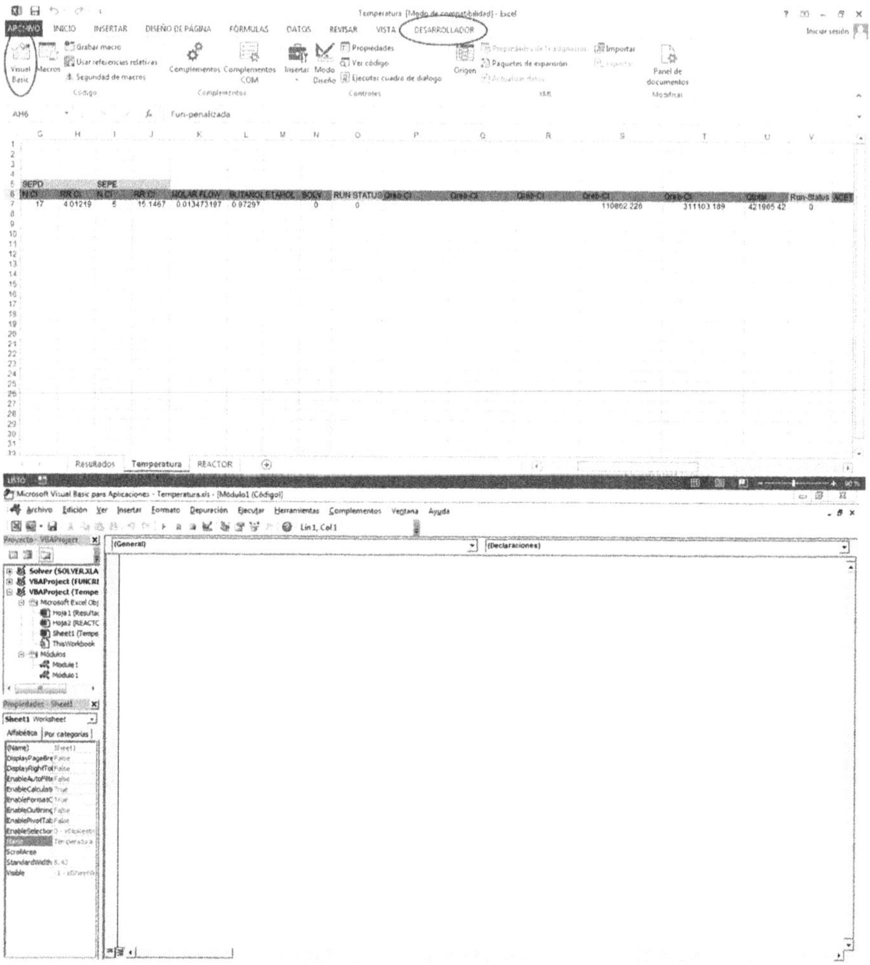

FIGURE 6.3
Visual basic inside Microsoft Excel.

Aspen simulation should be in a block named SEPD. In this example, we vary four variables (degrees of freedom): two variables on each distillation column. Figure 6.4 shows the relation of these four variables in the Aspen file.

Each variable in Aspen has its own route: all these routes could be located in the variable explorer of Aspen Plus. Figure 6.5 shows where it is possible to find all these routes. All screen captures were obtained in Aspen Plus V8. Similarly, the commands for the output data in Aspen Plus should be written, considering this Aspen Plus route, as shown in the following example:

```
Range("Temperature!K7").Select
ActiveCell.Value=Aspen.Tree.Data.Streams.WATERB.Output.
  MOLEFLOW.MIXED.BUTANOL.Value
```

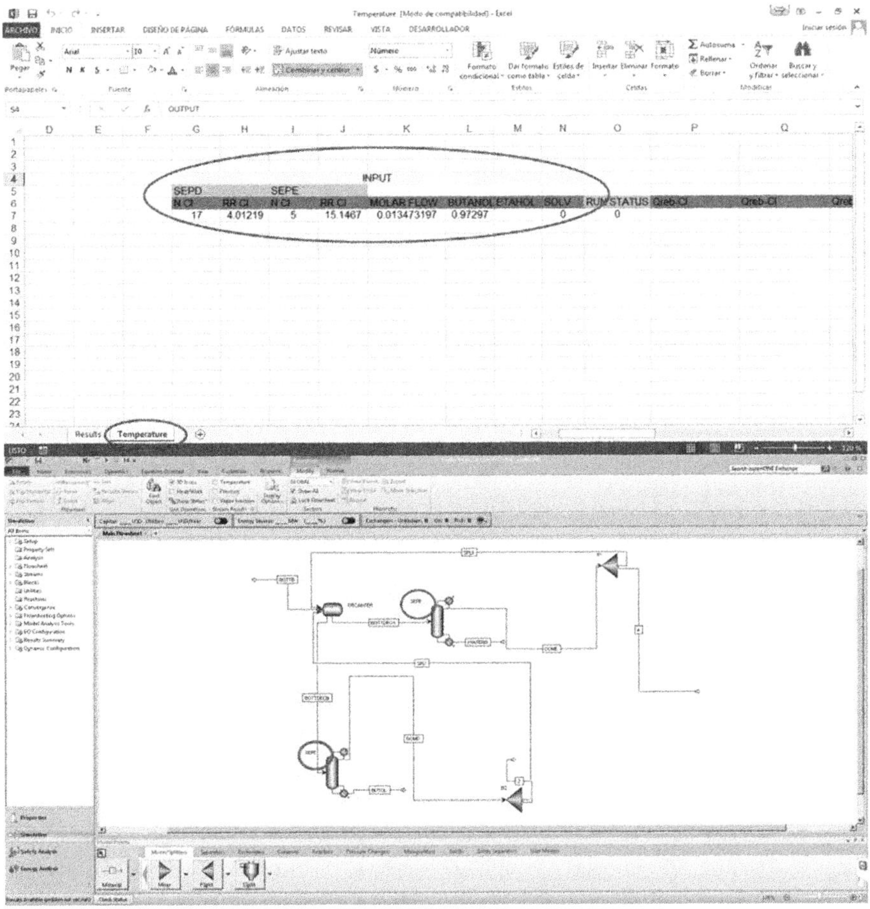

FIGURE 6.4
Input data in Microsoft Excel and Aspen Plus.

It indicates that the mole flow of butanol in the stream "WATERB" should be written in cell K7, in the sheet "Temperature.". The route could also found it in the same variable explorer. In this case, we need numerous flows, mass fraction, and reboiler heat duty of the two distillation columns as output data.

Once all the variables inside Aspen Plus are filled, the simulation is ready to run with the help of the command "Aspen.Engine.Run." Further, special attention must be paid to the following line:

```
Range("Temperature! O7").Select
ActiveCell.Value=Aspen.Tree.FindNode("\Data\ResultsSummary\
   RunStatus\Output\PER_ERROR").Value
```

This route defines whether the simulation runs with errors or with available results. Commonly, this route takes the value 0 for results available and

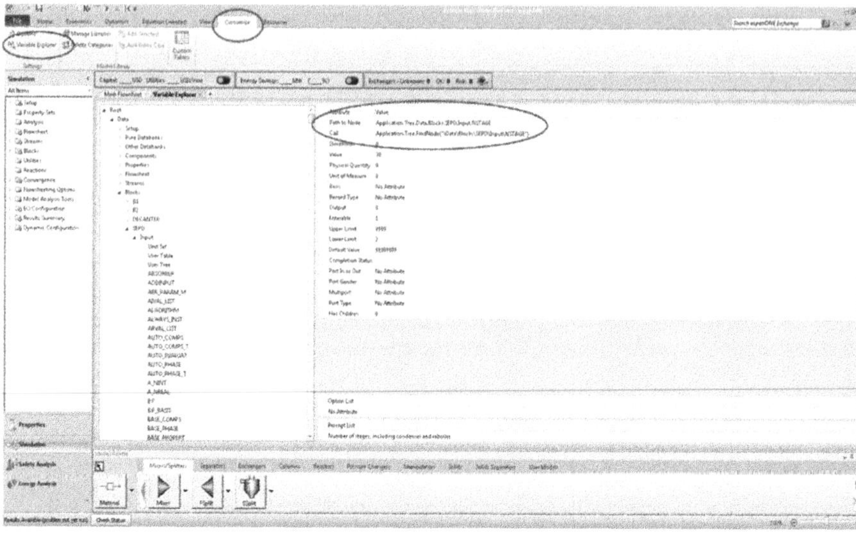

FIGURE 6.5
Variable explorer in Aspen Plus.

the value 1 for results available with errors. Consequently, it is up to the user whether this simulation is saved. The following "If" cycle helps us to save any simulation with 0 as value in the run status:

```
If ER = 0 Then
TempFileName = Range("Results!A1").Value
Aspen.SaveAs Filename:=TempFileName & ".bkp"
End If
```

We have already written all input and necessary output data. However, after a simulation, how the stochastic method will differentiate if a simulation is a good one? To discern that, it is necessary to consider some constrains. A simple way to write any constrain is described as follows:

```
=If(K7-0.0007<=0,0,(K7-0.0007)^100)
```

In this constrain, with the help of the conditioning "If," we compare the written data after simulation in cell K7. Therefore, this indicates that with this constrain we need a value less than 0.0007. However, if we did not achieve the required value, then the difference of the achieved value and 0.0007 will be multiplied by 100. It is recommended one constrain by one output data. Theoretically, if we have good limits for all variables, we are ready to accomplish all constrains. We recommend introducing a cell to add all those constrains. In this manner, it is possible to know the improvement of the method iteration after iteration. Finally, we consider the cell AH7 as the objective function, which is the total annual cost of both distillation columns. Nevertheless, any function inside Microsoft Excel can be calculated and can be taken as the objective function. For example, we can calculate the environmental impact of both distillation columns, likely with the help of the reboiler heat duty in output data. Therefore, final results of this impact could be displayed in any cell and considered as the objective function.

Furthermore, it is also possible to add any constrain in this cell (objective function); in other words, it is possible to differentiate a good simulation if all previous constrains are accomplished. For example, the conditioning =If(AF7=0,U7,U7+10000) could be used in cell AH7, which indicates that if summation (cell AF7) of constrains is zero, the value of U7 is written (total annual cost). Otherwise, the value plus 10000 is written to let the algorithm know that this is not a good vector.

In the last part of the code, we can find several lines starting with the following:

```
If ActiveSheet.Range("C1").Value >= 0 Then
    Range("A7:AV7").Copy
    Sheets("Results").Select
```

All these lines indicate that all results obtained in cells A7 to AV7 will be copied in the sheet "Results." In this manner, it is easy to analyze all vectors proposed in stochastic algorithm and to discern if we are indeed operating with good variable limits or not. It is necessary to start the counting with 3 in the first row as shown in Figure 6.6.

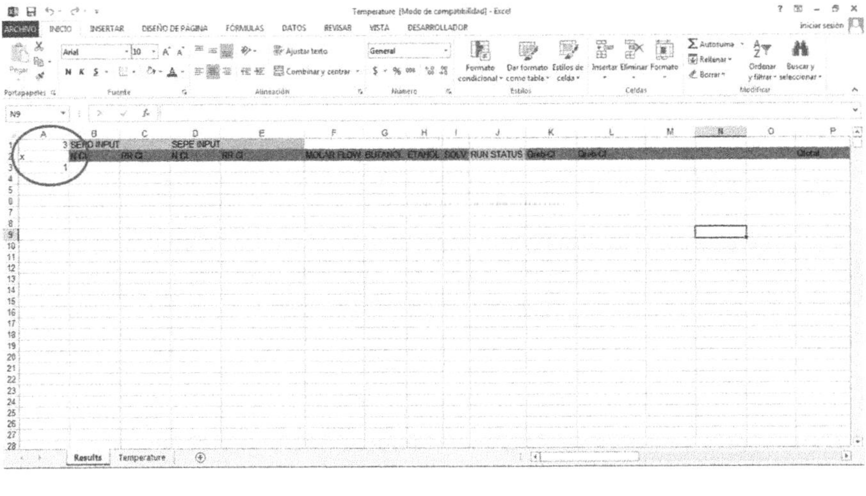

FIGURE 6.6
Sheet "Results" in Microsoft Excel.

6.4 Mono-Objective Optimization of a Multicomponent Distillation Column

At this point, we have developed all necessary codes to start the optimization. MATLAB is the main program, and the optimization toolbox will be used to optimize the Aspen Plus design. As an example, we consider simulated annealing as the mono-objective stochastic method. Once the optimization toolbox is open, it is necessary to select this stochastic method where it says "solver." Second, it is necessary to write the function to be optimized. In this case, we should write "@funobj" in MATLAB, which is linked with a subroutine in Microsoft Excel, as shown in Figure 6.7.

Finally, the starting point should be specified. In this example, we are working with four variables. Therefore, it is essential to define the starting point and boundaries as a vector of 4 by 1 and write as follows: [0.1 0.3 0.5 0.6]. The values are among 0 and 1 because we are working with normalized values.

The optimization process is started by just clicking the "start" button. We recommend to use at least one stopping criteria: one of the six options in the menu on the right side. Furthermore, at the end of this section, we also recommend to select any option about "plot function." In this manner, it is easy to achieve the best point obtained so far, the progress level of the optimization, etc. Furthermore, some annealing parameters can be selected, which can be tuned up to perform a faster and more efficient

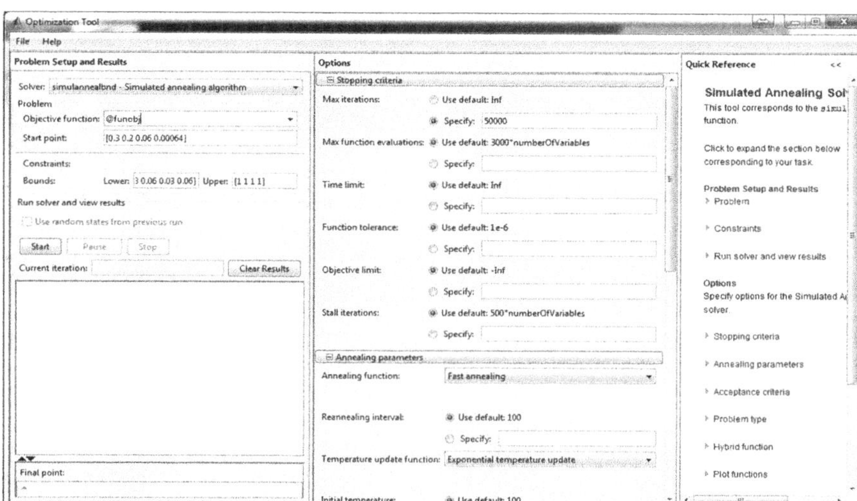

FIGURE 6.7
Using simulated annealing in optimization toolbox.

optimization process. Any doubt about this option could be clarified using MATLAB.

Considering a successful optimization, in the sheet "Results," the entire row will be copied, including the initialization vector declared as input data, the results, run status, output data, and objective function. Consequently, it is easy to analyze the progress of the objective function with regard to iterations. Figure 6.8 shows a complete optimization process and also the classical progress of optimization using simulated annealing algorithm. It also shows a simple way to represent the advance of the optimization process, considering the objective function. As recommended, a good stopping criterion is to compare the objective function obtained in the last iteration. If any improvement is not observed in several iterations, then it could be assumed that simulated annealing achieved the convergence at the tested numerical conditions.

6.5 Multi-Objective Optimization of a Multicomponent Distillation Column

A clear tendency to evaluate any process, from a wider point of view, involves the inclusion of some targets. Several authors (Segovia-Hernández et al., 2015; Sánchez-Ramírez et al., 2016) proposed to include economic and environmental targets; however, the inclusion of any target depends totally on the industrial necessity or any type of application.

This section briefly explains the application of the methodology learned so far, but considering at least two objective functions.

The entire methodology could be considered quite similar; however, before starting the process, one line of the MATLAB code must be modified. The last line of this code refers to the objective function. In multi-objective function, this line will slightly differ. The entire line is described as follows:

```
f=xlsread('C:\Users\PC\Documents\Temperature.xls', new,'Temper
  ature','Temperature', 'AH7:AI7');
```

As mentioned earlier, the main difference is in the last part. Now, we consider the two cells, AH7 and AI7, as objective functions. In this example, we consider two targets: the total annual cost and the environmental impact measured through eco-indicator 99. However, it may include more targets simultaneously, according to our requirements. Obviously, all those targets should be calculated inside the sheet "Temperature" in Microsoft Excel.

Once this small change is done, the optimization process is ready to start. In this example, we consider multi-objective optimization, using genetic

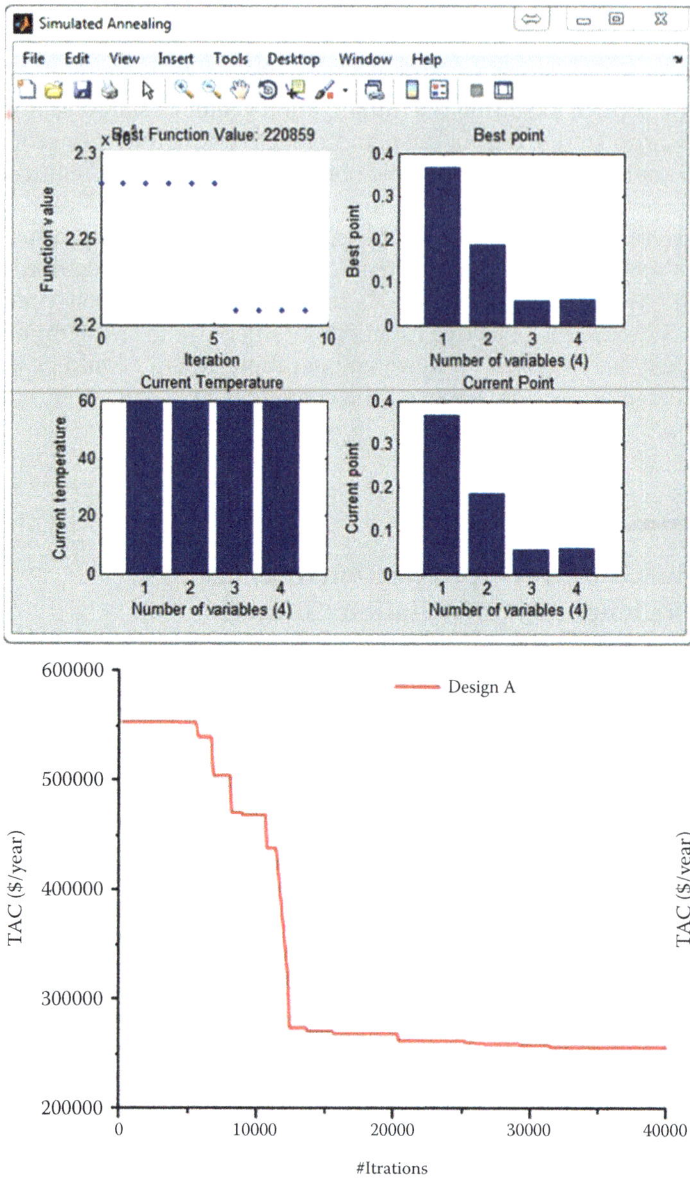

FIGURE 6.8
Results after optimization.

algorithms as the multi-objective genetic solver. This solver is included in the same optimization toolbox we have used so far.

In Figure 6.9, the display of this solver is quite similar to that we have used in mono-objective optimization. However, because we are considering

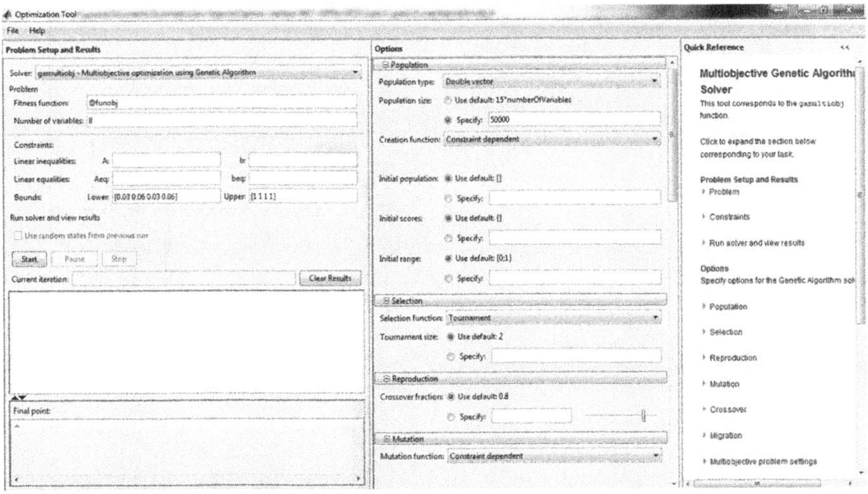

FIGURE 6.9
Multi-objective optimization using genetic algorithm.

a nature-inspired algorithm, the optimization process includes some other concepts, such as population size, selection, mutation, and crossover. Theoretically, all these parameters should be tuned up. However, some works have used this methodology (Gutiérrez-Antonio et al., 2009) with their own parameters. Therefore, we can use those data as base to perform a tune-up procedure. Although the solver asks constrains, it is not necessary to fill those requirements in the solver, as we are considering constrains inside Microsoft Excel.

The lower and upper boundaries should be filled similar to that done in mono-objective optimization. Moreover, the options in the menu on the right side (population, selection, reproduction, mutation, etc.) are already selected with default options. However, it can be changed according to the tune-up process.

Furthermore, we recommend to active at least one plot function. Our particular point of view is that the Pareto front is the most important option because it faces the two objective functions evaluated in this optimization process.

To start the optimization process, the user should just click the start button and wait until the procedure is complete. After successfully completing the optimization process, the results can be represented as a Pareto front, as shown in Figure 6.10.

The greater the amount of iteration, the better the shape of the Pareto front. However, with the increase in the amount of iterations, the optimization time and CPU effort will also increase. Therefore, this decision depends totally on the user. Moreover, every single point in the Pareto front represents a unique optimized design and all together represents the best solution achieved so far.

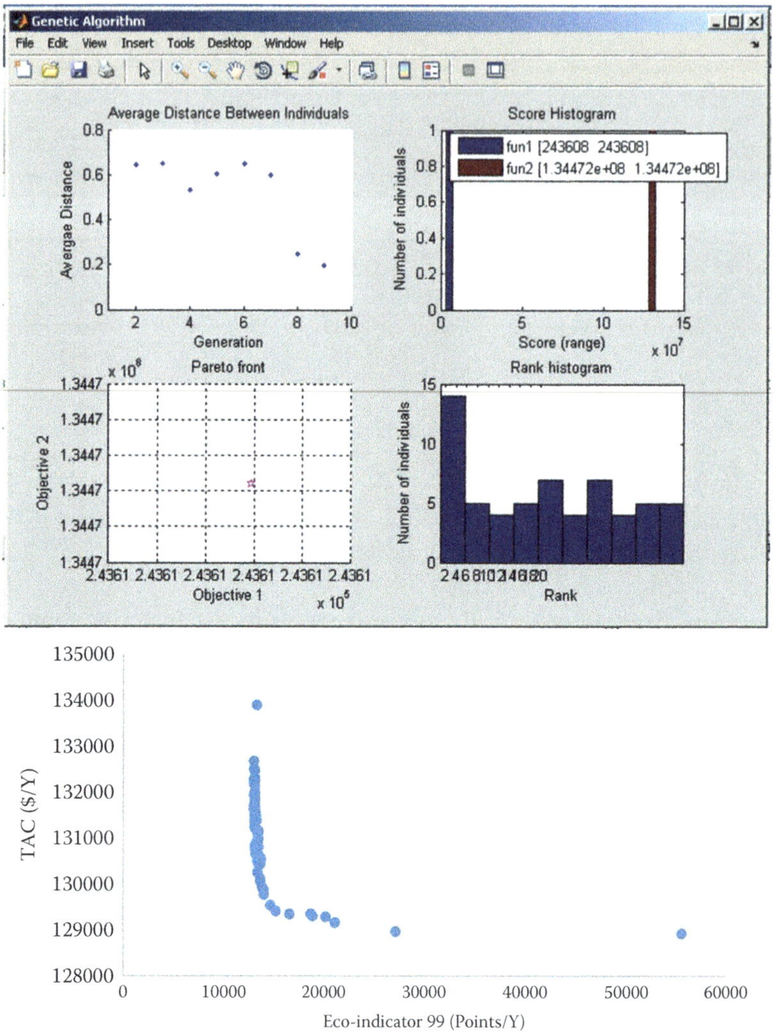

FIGURE 6.10
Pareto front facing the total annual cost and the eco-indicator 99.

6.6 Conclusions

In this chapter, using MATLAB as main program, a methodology to optimize two distillation columns has been described in detail, where numerous solvers can be used through the optimization toolbox. Also, it has been described how MATLAB uses Microsoft Excel as the handler of vectors between MATLAB and the model to evaluate the results with Aspen Plus. This robust methodology allows us to improve any model designed previously in Aspen Plus. In this chapter, we have considered two distillation columns as example; however, this methodology functions well for any model inside Aspen Plus. In our study, we have used the total annual cost and eco-indicator 99 as objective functions; however, any target can be considered. It could be simple or complex, as this type of stochastic algorithm only considers the input and output data.

References

C. Gutiérrez-Antonio, A. Briones-Ramírez, 2009, Pareto front of ideal Petlyuk sequences using a multiobjective genetic algorithm with constraints, *Comput. Chem. Eng.*, 33(2), 454–464.

M.S. Li, Q.H. Wu, T.Y. Ji, H. Rao, 2014, Stochastic multi-objective optimization for economic-emission dispatch with uncertain wind power and distributed loads, *Electr. Pow. Syst. Res.*, 116, 367–373.

E. Sánchez-Ramírez, J.J. Quiroz-Ramírez, J.G. Segovia-Hernández, S. Hernández, J.M. Ponce-Ortega, 2016, Economic and environmental optimization of the biobutanol purification process, *Clean Technol. Environ. Policy*, 18(2), 395–411.

J.G. Segovia-Hernández, S. Hernández, A.B. Petriciolet, 2015, Reactive distillation: A review of optimal design using deterministic and stochastic techniques, *Chem. Eng. Process*, 97, 134–143.

L. Wang, C. Singh, 2007, Environmental/economic power dispatch using a fuzzified multi-objective particle swarm optimization algorithm, *Electr. Pow. Syst. Res.*, 77(12), 1654–1664.

7

*Using External User-Defined Block Model in Aspen Plus®**

7.1 Introduction

This chapter presents a detailed procedure to extend the capabilities of Aspen Plus through the user block model. As described in Chapter 6, Aspen Plus is a user-friendly platform that allows the user to work with several programs. In this case, we use those capabilities to increase the models already available in Aspen Plus using MATLAB and Microsoft Excel to model a membrane unit that is used to separate an effluent coming from a reactor, mainly composed of biobutanol, ethanol, acetone, and water.

7.2 Importance of User-Defined Block Models

Nowadays, a very useful tool in several industrial sectors is Aspen Plus. Aspen Plus, using its models, allows the user to simulate industrial process (biochemical, polymer, specialty, etc.). Even in more recent versions, Aspen Plus has also considered economic and safety indicators.

However, despite Aspen Plus being a powerful tool, those built-in models already provided by the software probably do not fulfill the needs of the user. In this scenario, it is necessary to write a model inside Aspen Plus to reproduce a complete unit operation model. Further, the user might also want to write a model for physical properties, sizing, and costing. Moreover, special stream reports may be displayed if necessary. All these possibilities can be developed using the user model provided by Aspen Plus.

Aspen Plus provides four possibilities of the user model: *User*, *User2*, *User3*, and *Hierarchy*. Briefly, *User* helps to model any unit operation using a FORTRAN subroutine, which is the model to calculate the outlet streams produced in this model, if those outlet streams are based on model parameters and inlet streams. The limitation of *User* is that it allows only a maximum of four streams and one work/heat inlet stream. *User2* works similar to *User*; however, *User2* has no limits on the number of stream materials. Furthermore,

* Eduardo Sánchez-Ramírez, Juan José Quiroz-Ramírez, and César Ramírez-Márquez also contributed to the work on this chapter.

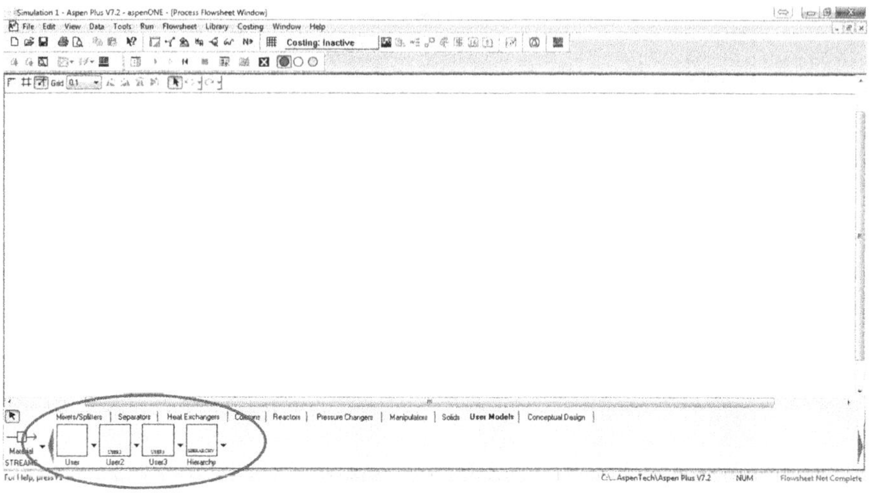

FIGURE 7.1
User model in Aspen Plus.

it is possible to develop all calculations through other software, such as Microsoft Excel. *User3* models are often used to simulate old built-in models, for example, RT-Opt and Aspen EO models from PML (Process Model Library). Finally, *Hierarchy* is used to create hierarchical structures pretty common in complex structures and complex schemes. Figure 7.1 shows where it is possible to locate the user models in Aspen Plus. Note all Aspen Plus screen captions are created using Aspen Plus V7.2.

As mentioned previously, the user model in Aspen Plus is always used to increase the capabilities of Aspen Plus and to develop any model according to our needs. Several reports show how the user model increases the scope of models; for example, Bao et al. (2002) have used this Aspen Plus option to simulate the industrial catalytic-distillation process in the production of methyl *tert*-butyl ether using experimental data. In this chapter, we consider a simplified production of biobutanol using a stoichiometric reactor followed by a membrane, which will be approached as the user model. This example certainly does not represent a real solution for the purification of an effluent coming from fermenter; however, it helps the user to learn and understand the methodology to consider a user model in Aspen Plus and the manner to link this user model with some external software such as MATLAB and Microsoft Excel.

7.3 Previous Work and Loading a User-Defined Block Model in Aspen Plus

As mentioned earlier, in this chapter, a membrane unit is considered as the first separation unit of an effluent coming from a fermentation reactor, which produces mainly acetone, butanol, and ethanol. Therefore, the first step is to consider the production of butanol. The production of this component is achieved using a stoichiometric reactor, which represents a great simplification of the real situation. However, because the main purpose of this chapter is to introduce the user model in Aspen Plus, we consider that it is possible to simplify this complex model of biobutanol production for something more didactic. Figure 7.2 shows the location of the reactors in Aspen Plus, for selecting a stoichiometric reactor. A stoichiometric reactor must be considered when reaction kinetic is unknown or unimportant; only the stoichiometry is known; and the user alone can specify the extent of reaction or conversion, and may consider the heat of reaction.

Once we select this reaction unit, it is necessary to introduce all components that are considered in the reaction. In this case, we consider acetone, *n*-butanol, ethanol, water, *n*-butyric acid, carbon dioxide, hydrogen, and dextrose (glucose). Further, we consider UNIFAC as thermodynamic model and an input flow stream of 1000 lb·mol/h of dextrose at 30°C and vapor fraction zero.

The complete set of stoichiometric reactions considered is described as follows (Gapes, 2000):

$$C_6H_{12}O_6 \rightarrow C_3H_6O + 3CO_2 + 4H_2 \tag{7.1}$$

$$C_6H_{12}O_6 \rightarrow C_4H_{10}O + 2CO_2 + H_2O \tag{7.2}$$

$$C_6H_{12}O_6 \rightarrow 2C_2H_6O + 2CO_2 \tag{7.3}$$

$$C_6H_{12}O_6 \rightarrow C_4H_8O_2 + 2CO_2 + 2H_2 \tag{7.4}$$

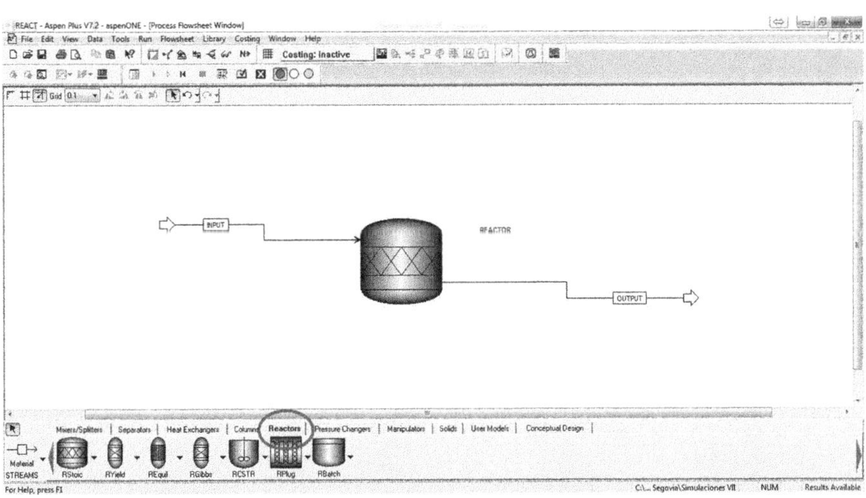

FIGURE 7.2
Stoichiometric reactor in Aspen Plus.

All these reactions are performed at 14.7 psia and 35°C. To introduce these reaction parameters, it is necessary to double click in the reactor and to write the data in "Specifications" (see Figure 7.3).

Also, all these reactions must be introduced one by one in the section "Reaction." Initially, the user must click "new." Then, a window is displayed, as shown in Figure 7.4.

In this window, the user must select all reactants and products, and must also fill the cell of "coefficient." Automatically, Aspen Plus will

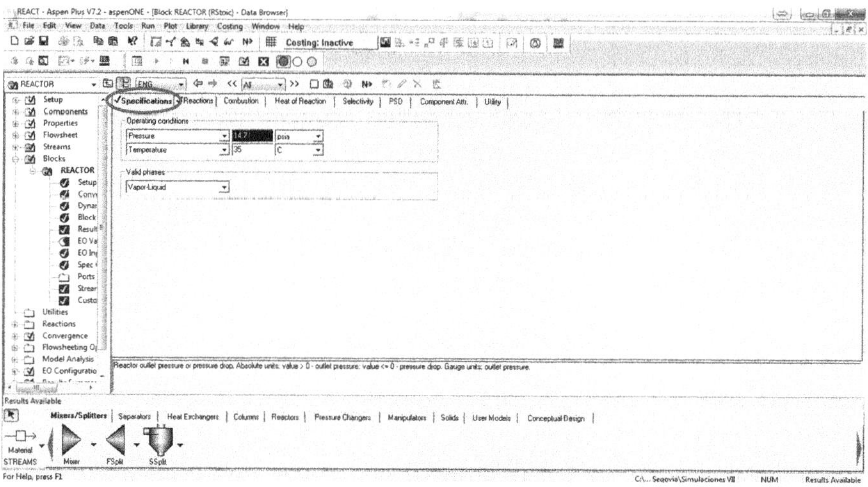

FIGURE 7.3
Reactor specification of a stoichiometric reactor.

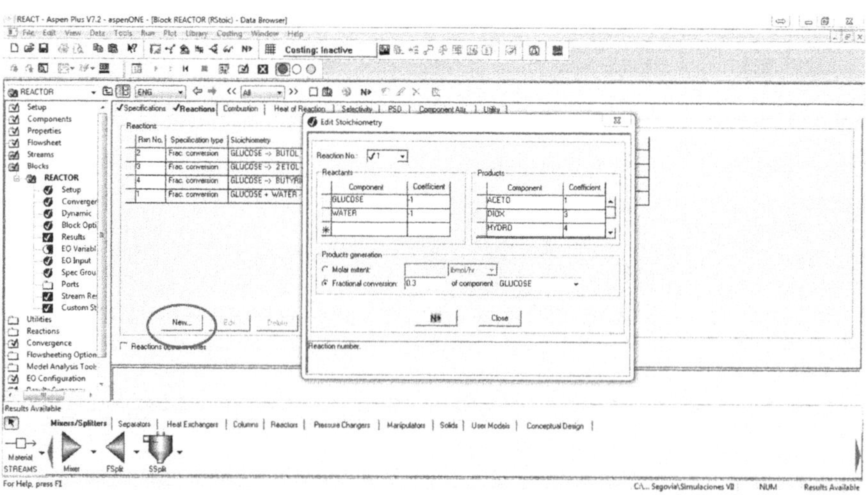

FIGURE 7.4
Stoichiometric reactions in Aspen Plus.

change the coefficients of reactants as negative number and those of products as positive number. Below this option, either "molar extent" or "fractional conversion" should be selected. In this case, we selected fractional conversion.

In this example, we consider a conversion of 0.3, 0.6, 0.098, and 0.002 for Equations 7.1 through 7.4, respectively. With all these options filled, we can run this stoichiometric reactor and can observe the stream and results, as shown in Figure 7.5.

Finally, to load a user model in this simulation alone, it is necessary to select "*User2*." Also, to connect the leaving stream of reactor to the input stream of "*User2*," click the right button and select "reconnect destination" to connect this stream to the module User 2. Furthermore, it is necessary to add two leaving streams, which represent the streams leaving the membrane (permeate and retentate). Note in this example, the outlet stream data are copied as input data for *User2*, see Figure 7.6.

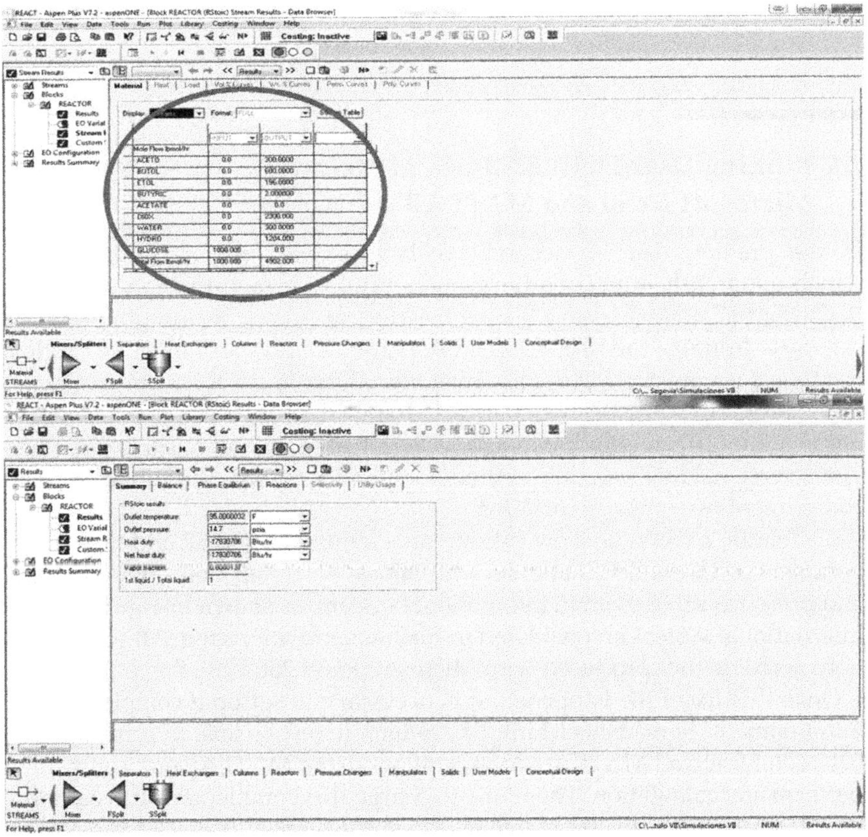

FIGURE 7.5
Stream and reactor results for producing butanol in a stoichiometric reactor.

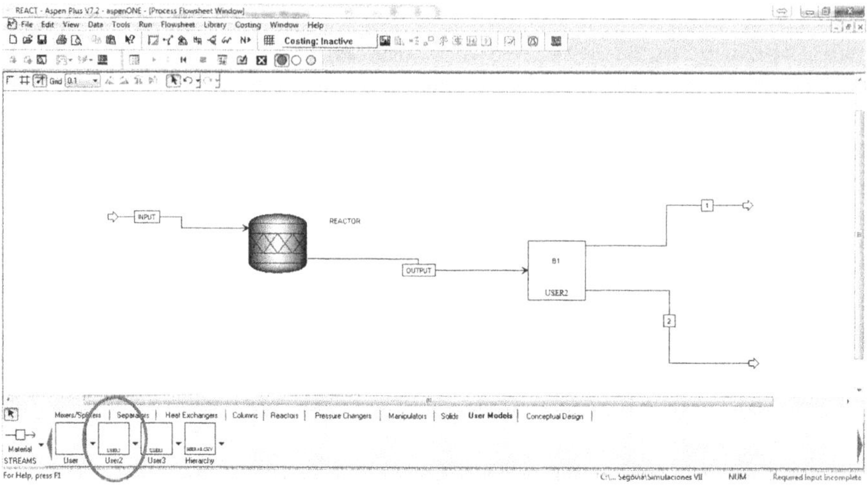

FIGURE 7.6
User2 in Aspen Plus.

7.4 Linking User-Defined Block Model with Microsoft Excel and MATLAB

In the previous section section, we have explained our case of study, which apparently would be the next to fulfill the requirements of Aspen; however, first, it is necessary to establish the link among the programs in this interface. Initially, the link between Aspen Plus and Excel must be created. For this, it is necessary to use an Excel file, which could be commonly found in C:\Program Files (x86)\AspenTech\Aspen Plus V7.2\Engine\User. It is possible that this methodology works better in Microsoft Excel 2007 or previous version. However, this methodology will work in a better manner if all compliments are already installed.

By default, Excel will show two streams named S1 and S2 and three components C1, C2, and C3, and several values of stream flows, temperature, and pressure, all of them in international system, as shown in Figure 7.7. The international system is considered in further simulations too. All screen captions were further produced with Microsoft Excel 2003.

Once this Excel file is opened, it is necessary to set up a complement for Excel named "Spreadsheet Link EX," which is responsible to communicate between Excel and MATLAB. Also, it allows using MATLAB environment to perform any calculation. Therefore, to charge this complement, it is essential to go to tool menu in Excel and select complements. In the next menu, by selecting browse and finding the route C:\Program Files\MATLAB\R2013b\

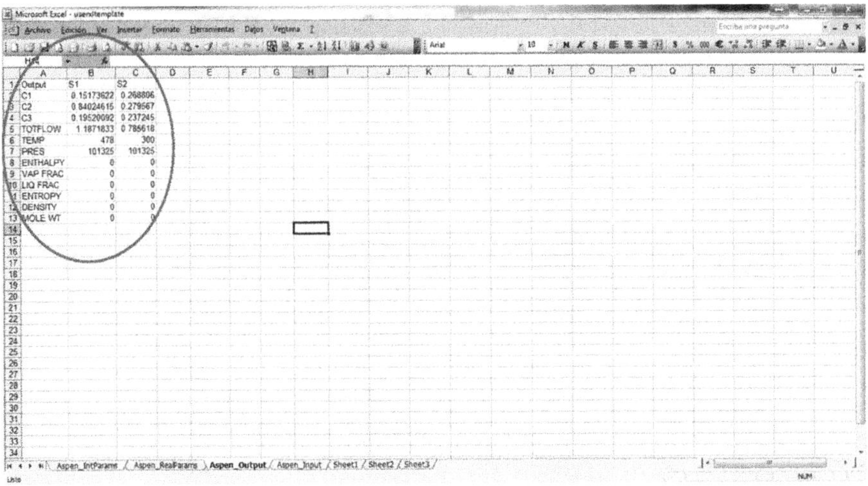

FIGURE 7.7
Default options in userxltemplate file.

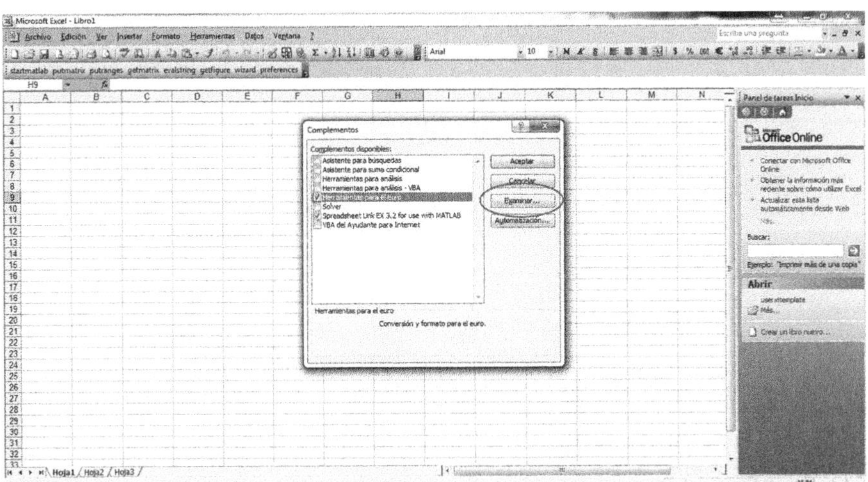

FIGURE 7.8
Adding Spreadsheet Link Ex 3.2 for use with MATLAB in Microsoft Excel.

toolbox\exlink (see Figure 7.8), the file excllink2003 can be found and selected. Once this complement is activated, MATLAB will open automatically. Moreover, the option Spreadsheet Link Ex 3.2 for use with MATLAB is selected.

MATLAB is now opened. Therefore, it is necessary to develop a function that allows us to solve some equations, which involve the mass balance of the two streams leaving the membrane (*User2*), permeate and retentate. The entire code of this function is as follows:

```
function
ParEnt,ParReal,CorSal]=usermod
   el_Pervaporador_EtOH(ParEnt,ParReal,CorEnt)
% This model calculates the retentate and permeate
compositions and flows
    k = ParEnt(8);
    xkf = ParReal(8);
    tam = size(CorEnt);
    n = tam(1) - 9;
    aux1 = [5 13];       s={};
    for i=1:n

        [pv(i,:),aux2,aux3,aux4,aux5,aux6]=propiedades(aux1(i));
        for g=1:length(aux3)
            g_secund(g,i)=aux3(g);
            numero(g,i)=aux4(g);
        end
        for g=1:length(aux5)
            g_secundliq(g,i)=aux5(g);
            numeroliq(g,i)=aux6(g);
        end
        s{i}=aux2;
    end
    Retentate = CorEnt(1:n,1)';
    switch tam(2)
        case 2
            Permeate(1,1:n) = CorEnt(1:n,2)';
        case 1
            Permeate(1,1:n) = zeros(1,n);
    end
    Pret = CorEnt(n+3,1)/1e5;
    Pperm = ParReal(4);
    tR = CorEnt(n+2,1);
    dA = ParReal(1);
    Perm = zeros(1,n);
    for i = 1:n
       Perm(i) = ParReal(4+i);
    end
    x(1,:) = CorEnt(1:n,1)'./CorEnt(n+1,1);
    N = [0 0];
    switch tam(2)
        case 2
            y(1,:) = CorEnt(1:n,2)'./CorEnt(n+1,2);
        case 1
            [pt,pvap] = presiondevapor(tR,pv,x(1,:));
            clear pt
```

```
            pvap = pvap'/0.9869;
            N = Perm.*x(1,:).*pvap;
            y(1,:) = N/sum(N);
    end
    top = x(1,k);
    c = 0;
    while top > xkf && min(N) >= 0
        c = c + 1;
        [pt,pvap] = presiondevapor(tR,pv,y(c,:));
        clear pt
        pvap = pvap'/0.9869;
        N = Perm.*(x(c,:).*pvap - y(c,:)*Pperm);
        Retentate(c+1,:) = Retentate(c,:) - (N/3600)* dA;
        Permeate(c+1,:) = Permeate(c,:) + (N/3600)*dA;
        x(c+1,:) = Retentate(c+1,:)/sum(Retentate(c+1,:));
        y(c+1,:) = Permeate(c+1,:)/sum(Permeate(c+1,:));
        top = x(c+1,k);     end
    ParEnt(1) = c;
    ParReal(3) = c*dA;
    switch tam(2)
        case 2
            CorSal = CorEnt;
        case 1
            CorSal(:,1) = CorEnt(:,1);
            CorSal(:,2) = CorEnt(:,1);
    end
    CorSal(n+3,2) = Pperm * 1e5;
    CorSal(1:n,1) = Retentate(end,:)';
    CorSal(1:n,2) = Permeate(end,:)';
    CorSal(n+1,1) = sum(Retentate(end,:));
    CorSal(n+1,2) = sum(Permeate(end,:));
end
```

Making a brief description of this code, in liquid phase, we consider an ideal behavior, which means that activity coefficients are equal to one. Also, we did not consider any change in temperature and pressure. The vector ParEnt contains X elements: ParEnt(1) is the number of calculated differential elements, and ParEnt(2) is the position of the component for which the final molar fraction is known. The vector ParReal contains eight elements: the size of a differential membrane area, the final molar fraction in the retentate of component, the membrane area, the permeate pressure, and the respective permeance of all components. The cycle stops only if either the molar fraction of component *k* is reached or if any flux is negative. This code should be saved in the same folder, where the work is performed.

The next step is to create a code (Macro) inside Visual Basic, which will be responsible for sending data from Aspen Plus to MATLAB. To do this, the option "macro" in the tools menu should be selected, as shown in Figure 7.9. Any name can be used for the macro created. In this example, "Macro1" is

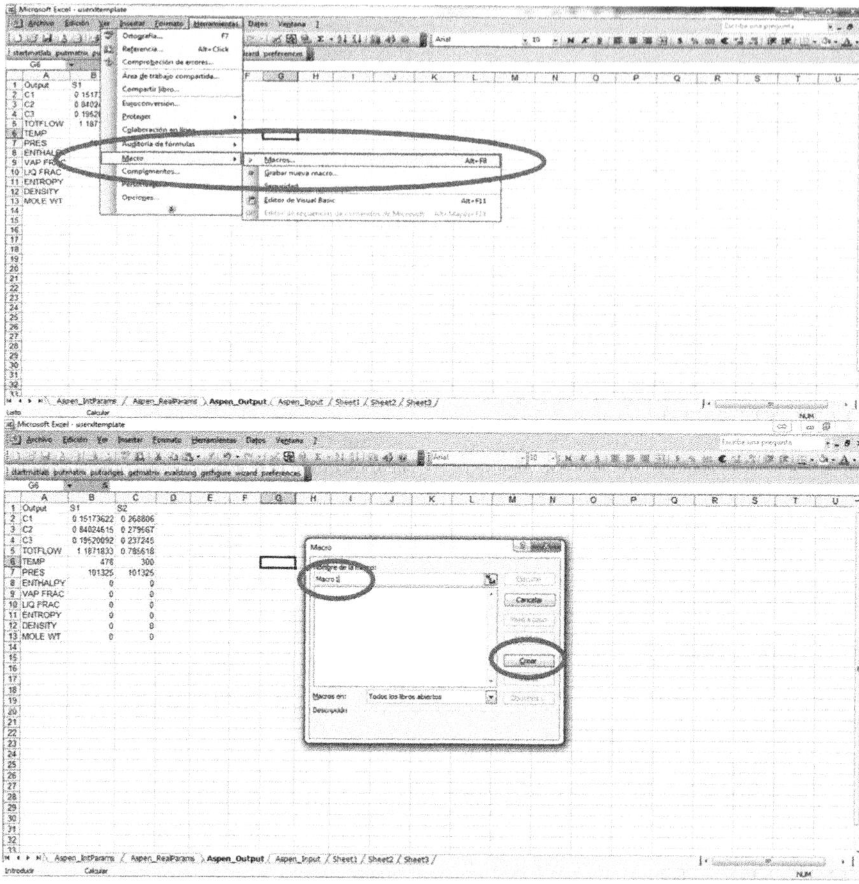

FIGURE 7.9
Creating a macro in Microsoft Excel.

used. Once we confirm that this macro is created, "Macro1" will be visible in the tools menu.

Now, following the same instructions in tools menu, we can modify this macro to write the necessary code for our purpose, see Figure 7.10.
The code in macro is written as follows:

```
Sub Macro1()
' Macro
    Sheets("Sheet1").Select
    Range("B36").Select
    ActiveCell.FormulaR1C1 = "=MLputMatrix(""ParEnt"",Aspen_
       IntParams)"
    Range("B37").Select
    ActiveCell.FormulaR1C1 = "=MLputMatrix(""ParReal"",Aspen_
       RealParams)"
```

```
    Range("B38").Select
    ActiveCell.FormulaR1C1 = "=MLputMatrix(""CorEnt"",Aspen_
       Input)"
    Range("B39").Select
    ActiveCell.FormulaR1C1 = _

       "=MLEvalString(""[ParEnt,ParReal,CorSal]=usermodel_
       Pervaporador_EtOH(ParEnt,ParReal,CorEnt)"")"
    Range("B40").Select
    ActiveCell.FormulaR1C1 = "=MLGetMatrix(""ParEnt"",""Aspen_
       IntParams"")"
    Range("B41").Select
    ActiveCell.FormulaR1C1 = "=MLGetMatrix(""ParReal"",""As
       pen_RealParams"")"
    Range("B42").Select
    ActiveCell.FormulaR1C1 = "=MLGetMatrix(""CorSal"",""As
       pen_Output"")"
    Range("B43").Select
    Sheets("Sheet1").Select
End Sub
```

It is essential to add three commands in the corresponding cells: MLPutMatrix, MLevalstring, and MLGetMatrix. MLPutMatrix has the responsibility of sending the matrix named "Aspen_RealParams" to MATLAB using the name "reales" and "C_entrada." MLevalstring is responsible for evaluating in Excel the function previously created in MATLAB, which has as input argument "c_entrada" and "reales" to obtain output streams and new values previously named "c_salida" and "realesc." MLGetMatrix is responsible for writing in Excel the matrix created by MATLAB named

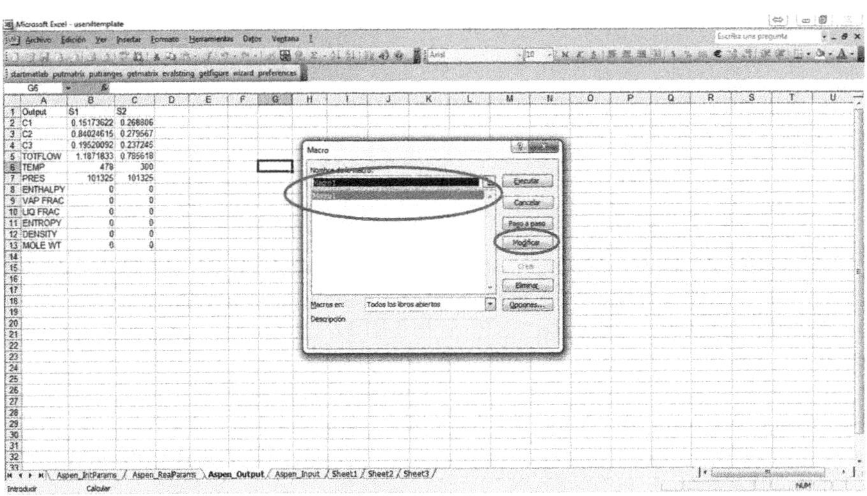

FIGURE 7.10
Modifying a macro in Microsoft Excel.

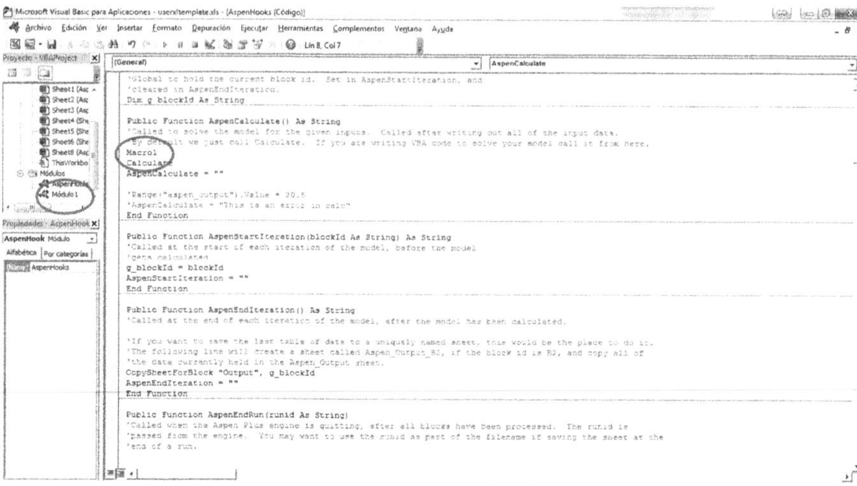

FIGURE 7.11
Adding a macro in Excel–Aspen file.

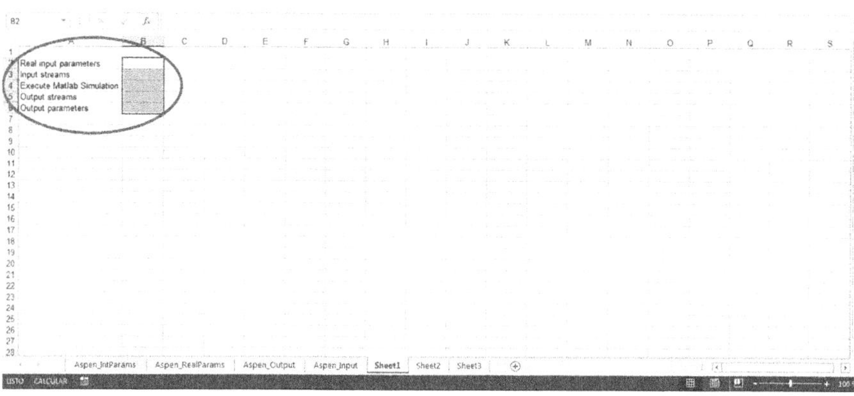

FIGURE 7.12
Adding a macro in Excel–Aspen file.

"C_salida," where the group of cells named "Aspen_Output" has been assigned. Furthermore, MLGetMatrix rewrites in Excel the matrix created in MATLAB named "realesc," where it will be assigned as a group of cells called "Aspen_RealParams."

This new macro must be included inside the macro previously developed by Aspen Plus to execute this Excel file in the user model. In this manner, when it is time to run Aspen Plus, both macros will be executed, as shown in Figure 7.11.

Now, when this macro is executed, zeroes will be inserted as values in the column for each command, as shown in Figure 7.12. Now, we are ready to

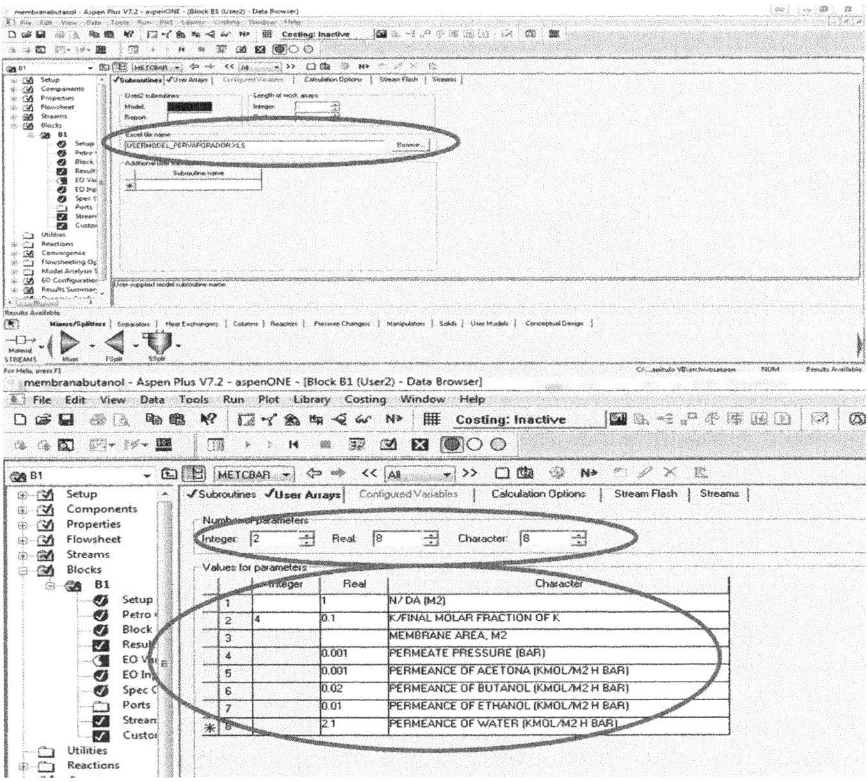

FIGURE 7.13
Filling up *User2* in Aspen Plus.

continue with our Aspen simulation. It is possible to change the name of the Excel file as required, and it is totally optional. The next step is to fulfill all requirements of the user model. Initially, it is necessary to search our Excel file by clicking the button "browse." Here, it is essential to select the correct path where the Excel file is located (see Figure 7.13). It is essential to leave both the options "*User2* subroutines" and "Length of work arrays" empty.

It is important to fill the section user arrays with the correct numbers. In this case, the option Integer, Real, and Character should be filled with 2, 8, and 8, respectively. These numbers correspond to the area of membrane and the position of water in the input matrix (component to remove), the number 8 corresponds to the input parameters of Aspen, named "Real" in MATLAB. In this Aspen simulation, the column with the name of each arrow is filled; however, this action is totally optional. This simulation can work without this text. Note also in the section below, it is essential to fill the respective values, which are mostly about permeances in kmol/m^2 of all species being investigated. According to the bibliography, those values can be changed.

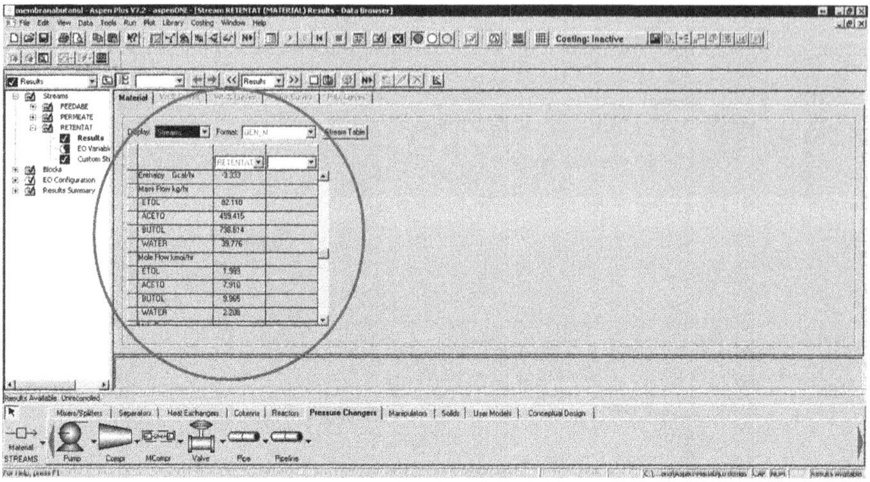

FIGURE 7.14
Streams results in *User2* after simulation.

Furthermore, the cell named "membrane area" is left empty because the results generated by MATLAB are stored here. Now, we are ready to run the simulation just by clicking next in Aspen Plus to obtain the results. Figure 7.14 shows the stream results obtained after simulation. Moreover, the retentate streams are also shown in Figure 7.14. However, it is possible to watch any stream. Now, with this methodology, we are ready to reproduce it in any case of study, just being careful regarding introducing the components to Aspen Plus because MATLAB does not understand about chemical components. MATLAB understands about arrows and columns; therefore, it is important to introduce the components in Aspen Plus in a correct order.

Furthermore, this methodology is highly improbable because we are considering ideal behavior; however, the MATLAB function can be programmed according to our requirements.

Besides, we are considering the flow coming from a reactor as input streams to *User2*; however, a distillation column or any other unit can be considered.

7.5 Conclusions

In this chapter, the creation of a user model in Aspen Plus has been achieved. Besides, we were able to link Microsoft Excel and MATLAB with Aspen Plus to turn this user model into a membrane that is selective to butanol. Initially, we simulated a reactor to create the feed stream to the membrane, which further separates this stream into permeate and retentate. Although this methodology was created to perform the work of a membrane, we can use this methodology to reproduce anything we want according to the needs of a designer or user.

References

J. Bao, B. Gao, X. Wu, M. Yoshimoto, K. Nakao, 2002, Simulation of industrial catalytic-distillation process for production of methyl tert-butyl ether by developing user's model on Aspen Plus platform, *Chem. Eng. J.*, 90(3), 253–266.

J.R. Gapes, 2000, The economics of acetone-butanol fermentation: Theoretical and market considerations, *J. Mol. Microbiol. Biotechnol.*, 2(1), 27–32.

8

*Optimization with a User Kinetic Model**

8.1 Introduction

Several intensified processes are found to be quite superior to the conventional technology by multiple already-published studies (Doherty and Buzzad, 1992). A simple example is reactive distillation, which presents several benefits over conventional distillation for performing some processes such as esterification, transesterification, alkylation, and the simultaneous synthesis and purification of other several compounds.

Aspen Plus® is a software that allows the user to model a distillation column involving a chemical reaction through the kinetic model included inside Aspen Plus. However, sometimes the chemical reaction is somehow complicated and cannot be reproduced with the tool provided by Aspen Plus. Under this scenario, Aspen Plus allows the user to load a subroutine inside an Aspen Module, which we call external kinetic model.

Therefore, this chapter introduces some generalities about using Aspen Plus modeler to design a process where it is necessary to handle a kinetic model. In particular, we approach this problem from the point of view of reactive distillation using two examples. The first case of study involves the chemical reaction from an external FORTRAN code linked with Aspen Plus to reproduce a more complex kinetic model, and the second case considers the chemical reaction from the inside of Aspen Plus.

8.2 Kinetic Models Allowed in Aspen Plus

Aspen Plus is a very useful tool to simulate and design industrial processes. However, sometimes the user requires a more specific model that probably is not included inside Aspen Plus. To satisfy those needs, Aspen Plus software includes a module named *User Model* which makes it possible to write a more complete code (Aspen Technology Inc., 2001).

For this, it is possible that the User Model options may be written from a FORTRAN subroutine. Therefore, it is important to have some knowledge about User Model and basic programming skills. As a brief description, Aspen

* Eduardo Sánchez-Ramírez and Juan José Quiroz-Ramírez also contributed to the work on this chapter.

Plus linked with FORTRAN by the User Model consists of one or more subroutines that can be written to extend the capabilities of Aspen Plus. It is possible to find six types of FORTRAN User models inside Aspen Plus (see Figure 8.1).

The User Model makes it possible to calculate and design some unit operations in Aspen Plus. Figure 8.2 provides some examples of such operations.

Each Aspen Plus Version works with certain FORTRAN versions and Microsoft Visual Studio. Therefore, it is necessary to review the user manual. Aspen Plus V8.4 is used throughout this chapter; therefore, it is essential to know all the requirements of Aspen Plus.

Aspen Plus V8.4 is based on the Intel Fortran 9.1 compiler and Microsoft Visual Studio 2008 SP1. This is different from version 2004.1 and earlier versions of Aspen Plus, which used Compaq Visual Fortran instead. Aspen Plus has moved away from Compaq Visual Fortran because this compiler is no longer supported by its manufacturer and may be difficult for some customers to obtain.

Aspen Plus provides a utility to allow the user to specify the combination of compiler and linker. This utility sets certain environment variables, so that the scripts mentioned in this manual and Aspen Plus will use the tools specified by the user to compile and link FORTRAN. When the user first installs Aspen

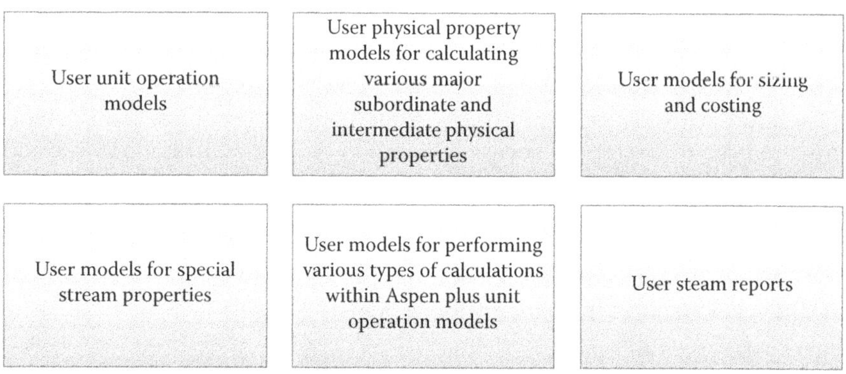

FIGURE 8.1
User model for use in Aspen Plus.

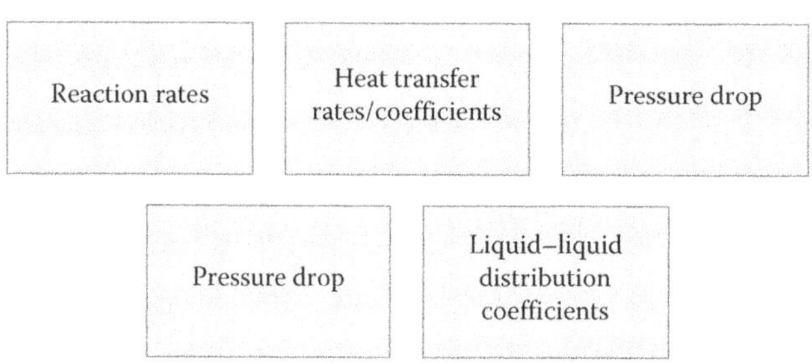

FIGURE 8.2
Possible unit operations to be modeled through the User Model.

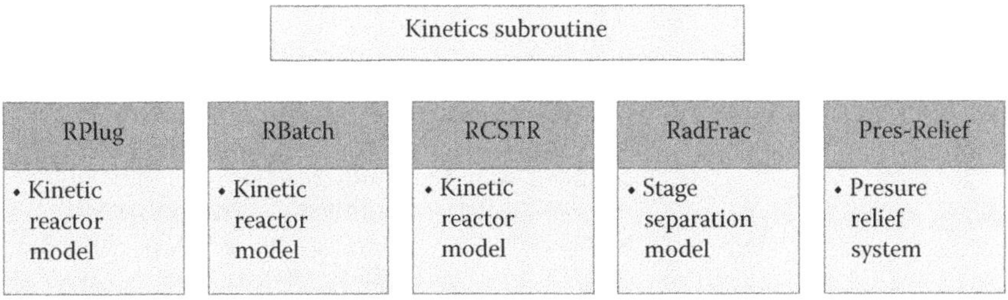

FIGURE 8.3
User Kinetics Subroutines in Aspen Plus.

Plus, the utility runs after rebooting. If the user wants to run it at any other time, it can be found under the Start menu under Programs | AspenTech | Process Modeling | Aspen Plus | Select Compiler for Aspen Plus.

User models written in FORTRAN should follow these rules and conventions: Filenames: Files may be given any name, but should end with a .f file extension. If the .for extension is selected, names beginning with an underscore (example.for) should not be used because Aspen Plus may overwrite these files with files containing non-interpretable inline FORTRAN from models such as Calculator blocks. If the user wants to call Aspen Plus, a kinetics subroutine should be supplied to calculate the reaction rates for the models as shown in Figure 8.3.

For RadFrac module, the user kinetics subroutine in either USER reaction type or REAC-DIST reaction type can be supplied. The kinetics subroutine calculates the rate of generation for each component in each stream. If solids participate in the reactions, the kinetics subroutine also accounts for changes in the outlet stream particle size distribution, and in the component attribute values.

For RadFrac module, the kinetics subroutine calculates the rate of generation for each component on a given stage. It also calculates the individual reaction rates in each phase if this routine is used for rate-based calculations.

8.3 Developing a User Kinetic Model

To develop a kinetic model in FORTRAN, which will be further used with Aspen Plus, it is necessary to write the entire code adequately, which means with a correct format and a correct variable arrangement. The lines shown in Table 8.1 show the main parts of a FORTRAN code, starting with the declaration of variables.

The meaning of each variable is denoted in Table 8.2. These variables may or may not be specified in the code to develop. It is essential to write all the code instructions for a particular purpose.

Almost all the arguments used in the subroutine are information required for the model that is taken from the simulation in Aspen Plus when the subroutine is called. Others are useful as an information to link the Aspen Plus simulation and the FORTRAN subroutine.

TABLE 8.1

Variables in FORTRAN Code

`SUBROUTINE subrname1*`	`(N, NCOMP, NR, NRL, NRV, T, TLIQ, TVAP, P, PHFRAC, F, X, Y, IDX, NBOPST, KDIAG, STOIC, IHLBAS, HLDLIQ, TIMLIQ, IHVBAS, HLDVAP, TIMVAP, NINT, INT, NREAL, REAL, RATES, RATEL, RATEV, NINTB, INTB, NREALB, REALB, NIWORK, IWORK, NWORK, WORK)`

TABLE 8.2

Variables in FORTRAN Code

Variable	I/O†	Type	Dimension	Description
N	I	INTEGER	–	Stage number
NCOMP	I	INTEGER	–	Number of components present
NR	I	INTEGER	–	Total number of kinetic reactions
NRV	I	INTEGER	3	NRL (1)—Number of overall liquid reactions NRL (2)—Number of liquid1 reactions NRL (3)—Number of liquid2 reactions
T	I	REAL*8	–	Number of vapor phase kinetic reactions
TLIQ	I	REAL*8	–	Stage temperature (K)
TVAP	I	REAL*8	–	Vapor temperature on stage (K)
P	I	REAL*8	–	Stage pressure (N/m^2)
PHFRAC	I	REAL*8	3	Phase fraction (1) Vapor fraction (2) Liquid1 fraction (RadFrac, 3-phase) (3) Liquid2 fraction (RadFrac, 3-phase)
F	I	REAL*8	–	Total flow on stage or segment (Vapor +Liquid) (kmol/s)
Y	I	REAL*8	NCOMP	Vapor mole fraction
IDX	I	INTEGER	NCOMP	Component sequence number
NBOPST	I	INTEGER	6	Property option set (see NBOPST)
KDIAG	I	INTEGER	–	Local diagnostic level
STOIC	I	REAL*8	NCOMP, NR	Reaction stoichiometry (see STOIC)
IHLBAS	I	INTEGER	–	Basis for liquid holdup specification 1: volume, 2: mass, 3: mole
HLDLIQ	I	REAL*8	–	Liquid holdup IHLBAS Units 1 m^3 2 kg 3 kmol

(Continued)

TABLE 8.2 (*Continued*)

Variables in FORTRAN Code

Variable	I/O†	Type	Dimension	Description
TIMLIQ	I	REAL*8	–	Liquid residence time (s)
IHVBAS	I	INTEGER	–	Basis for vapor holdup specification 1: volume, 2: mass, 3: mole
HLDVAP	I	REAL*8	–	Vapor holdup IHVBAS Units 1 m^3 2 kg 3 kmol
IHVBAS	I	REAL*8	–	Vapor residence time (s)
TIMVAP	I	INTEGER	–	Number of integer parameters (from Reactions Subroutine sheet)
NINT	I	INTEGER	–	Number of integer parameters (from Reactions Subroutine sheet)
INT	I/O	INTEGER	NINT	Vector of integer parameters (from Reactions Subroutine sheet)
NREAL	I	INTEGER	–	Number of real parameters (from Reactions Subroutine sheet
REAL	I/O	INTEGER	NREAL	Vector of real parameters (from Reactions Subroutine sheet)
RATES	O	REAL*8	NCOMP	Component reaction rates (kmol/s)
RATEL	O	REAL*8	NRL	Individual reaction rates in the liquid pha (kmol/s) (for Rate-Based Distillation)
RATEV	O	REAL*8	NRV	Individual reaction rates in the vapor phase (kmol/s) (for Rate-Based Distillation)
NINTB	I	INTEGER	–	Number of integer parameters (from unit operation block UserSubroutine form)
INTB	I/O	INTEGER	NINTB	Vector of integer parameters (from unit operation block UserSubroutine form)
NREALB	I	INTEGER	–	Number of real parameters (from unit operation block UserSubroutine form)
REALB	I/O	INTEGER	NREALB	Vector of real parameters (from unit operation block UserSubroutine form)
NIWORK	I	INTEGER	–	Length of integer work vector (from unit operation block UserSubroutine form)
IWORK	W	INTEGER	NIWORK	Integer work vector
NWORK	I	INTEGER	–	Length of real work vector (from unit operation block UserSubroutine form)
WORK	W	INTEGER	NIWORK	Real work vector

I is an input subroutine; O is an output from subroutine; and W is the workspace.

With this set of variables, it is possible to create a code according to the user requirements and also from the workspace of Aspen Plus. In this chapter, we approach a reactive distillation process using a kinetic external model; therefore, it is necessary a code as simple as possible to develop this task. For further information, see Section 8.4, where the example is described in detail. The executable code developed for the case of study of production of di-n-pentyl ether (DNPE) is presented as follows (Bildea et al., 2015):

```
C     calculation of DNPE synthesis reaction rate
C
C     kinetics taken from
C     Pera-Titus et al., Chemical Engineering and Processing,
         48(2009) 1072
C
C     original version: dnpe.f,
C                  - reaction rate in kmol/m3/s, assuming 1512
                      kg/m3
C
C-------------------------------------------------------------
C
C     User Kinetics Subroutine for RADFRAC, BATCHFRAC, RATEFRAC
C
      SUBROUTINE DNPECSB(NSTAGE, NCOMP,   NR,     NRL,    NRV,
      2        T,  TLIQ, TVAP, P,   VF,
      3        F,  X,    Y, IDX,  NBOPST,
      4        KDIAG, STOIC, IHLBAS,HLDLIQ, TIMLIQ,
      5        IHVBAS,HLDVAP, TIMVAP, NINT, INT,
      6        NREAL, REAL, RATES, RATEL,  RATEV,
      7        NINTB,INTB, NREALB, REALB,  NIWORK,
      8        IWORK, NWORK, WORK)
C       IMPLICIT NONE
C
C     DECLARE VARIABLES USED IN DIMENSIONING
C
      INTEGER NCOMP, NR, NRL, NRV, NINT,
      +   NINTB, NREALB,NIWORK,NWORK, N_COMP
C
C     DECLARE PARAMETERS & VARIABLES USED IN PARAMETERS
C
C     component order
C     ===============
C     this routine assumes that the components are in this order:

      INTEGER K_PENTANOL,K_H2O, K_DNPE
      PARAMETER(K_PENTANOL=1)
      PARAMETER(K_H2O=2)
      PARAMETER(K_DNPE=3)
      PARAMETER(N_COMP=3)
C-------------------------------------------------------------
```

```
C
C     DECLARE ARGUMENTS
C
      INTEGER IDX(NCOMP),    NBOPST(6),     INT(NINT),
     +         INTB(NINTB),   IWORK(NIWORK),NSTAGE,
     +         KDIAG, IHLBAS,IHVBAS,NREAL, KPHI,
     +         KER,    L_GAMMA,       J, K
      REAL*8 X(NCOMP,3),    Y(NCOMP),
     +        STOIC(NCOMP,NR),      RATES(NCOMP),
     +        RATEL(NRL),    RATEV(NRV),
     +        REALB(NREALB),WORK(NWORK),  B(1),   T,
     +        TLIQ,   TVAP,   P,      VF,     F
      REAL*8 HLDLIQ,TIMLIQ,HLDVAP,TIMVAP,TZERO,
     +        FT
C
C
C     DECLARE SYSTEM FUNCTIONS
C
      REAL*8 DLOG
C
C     DECLARE LOCAL VARIABLES
C
      INTEGER IMISS, IDBG
      REAL*8 REAL(NREAL),  RMISS, Keq, KW, k1, k01, ap, ae, aw,
     +            RATE(4), RATNET(4)
      REAL*8 PHI(N_COMP)
      REAL*8 DPHI(N_COMP)
      REAL*8 ACTIV(N_COMP)
C
#include "ppexec_user.cmn"
      EQUIVALENCE (RMISS, USER_RUMISS)
      EQUIVALENCE (IMISS, USER_IUMISS)
C
C
#include "dms_maxwrt.cmn"
#include "dms_ipoff3.cmn"
#include "dms_lclist.cmn"
      INTEGER FN
#include "dms_plex.cmn"
      EQUIVALENCE(B(1),IB(1))
      FN(J)=J+LCLIST_LBLCLIST
C
#include "dms_rglob.cmn"
C
C      DATA STATEMENTS
C
      DATA IDBG/0/
C      thermodynamic rate constant DKA
C      ===============================
 9010 FORMAT(1X,3(G13.6,1X))
```

```
 9000 FORMAT(' fugly failed at T=',G12.5,' P=',G12.5,' ker=',I4)
 9020 FORMAT(' compo ',I3,' mole-frac=',G12.5,' activity=',G12.5)
 9030 FORMAT(' stage=',I4,' spec-rate=',G12.5,' net-rate=',G12.5)
C
C     BEGIN EXECUTABLE CODE
C
c     open(UNIT=1, FILE='test.txt')
C   Equilibrium constant
    Keq = 8.9229*DEXP(778.69/T)
C   Water adsorption constant
    KW = DEXP(1.46 - 6615*(1/T - 1.0/438.0))
 C Reaction rate constant, mol/h/kg_cat
    k1 = DEXP(2.808  - 11595*(1/T - 1.0/438.0))
 C Reaction rate constant, kmol/s/kg_cat
    k1 = k1/3600.0 / 1d3

C      reaction rate constant
C      ======================
C
       IF(IDBG.GE.1)THEN
         WRITE(MAXWRT_MAXBUF(1),9010) Keq, KW, k1, k01
         CALL DMS_WRTTRM(1)
       ENDIF

C      calculation of components activities
C      ====================================
C      calculate only fugacity coefficient
      KPHI=1
C      fugacity coefficient of components in the mixture
       CALL PPMON_FUGLY(T,P,X(1,1)
      +      ,Y,NCOMP,IDX,NBOPST,KDIAG,KPHI,PHI,DPHI,KER)
       IF(KER.NE.0)THEN
         WRITE(MAXWRT_MAXBUF(1),9000) T,P,KER
         CALL DMS_WRTTRM(1)
       ENDIF
C      set offset to get activity coefficients
C      (see vol5, p 11-11 and asp$sor search for 'GAMMAL')
      L_GAMMA=IPOFF3_IPOFF3(24)
C      calculate activities for plex data
      DO J=1,NCOMP
         ACTIV(J=DEXP(B(FN(L_GAMMA)+J))*X(J,1)
      END DO
C
       ap = ACTIV(K_PENTANOL)
       aw = ACTIV(K_H2O)
       ae = ACTIV(K_DNPE)
C
       IF(IDBG.GE.1)THEN
         DO J=1,NCOMP
           WRITE(MAXWRT_MAXBUF(1),9020) J,X(J,1),ACTIV(J)
```

```
           CALL DMS_WRTTRM(1)
         END DO
       ENDIF
C      reaction rate
C      =============
      RATE(1) = k1*(ap*ap - aw*ae/Keq) / ap / (1+KW*dsqrt(aw))
C
      DO K = 1,NRL
         RATE(K) = RATE(K) * HLDLIQ
      END DO
C      INITIALIZATION OF COMPONENT REACTION RATES
C
      DO J = 1,NCOMP
          RATES(J) = 0.D0
        END DO

C
C      COMPONENT REACTION RATES in kmol/sec
C
      DO K=1,NRL
       DO J=1,NCOMP
         IF (DABS(STOIC(J,K)) .GE. RGLOB_RMIN) RATES(J) = RATES(J) +
      1   STOIC(J,K) * RATE(K)
         END DO
       END DO

       IF(IDBG.GE.1)THEN
        WRITE(MAXWRT_MAXBUF(1),9030) NSTAGE,RATE(1)
        CALL DMS_WRTTRM(1)
       ENDIF

c       write(1,*) NSTAGE, k1, Keq, KW,
c       write(1,*) aw, ap, ae
c       write(1,*)
       RETURN
#undef P_MAX3
       END
```

Then, it is necessary to save this file as a FORTRAN file with the extension ".f" or ".for." Once the code is ready, it is possible to use it when the name of the file is introduced in the corresponding box in the subroutine form. Therefore, with all these attachments, it is possible to develop a link between Aspen Plus and FORTRAN. This entire methodology is further explained in Section 8.4.

8.4 Loading a User Kinetic Model in Aspen Plus

Once the FORTRAN code is written, it is mandatory to compile that code with a proper tool provided by Aspen Plus, the Aspen Plus Simulation

Engine. The goal of compiling process is to translate the code from the Visual Basic language to FORTRAN, obtaining two files with extension .obj and .dll. Figure 8.4 presents the entire process to establish a link between Aspen Plus and FORTRAN.

The versions of the programs required for the code development and compiling steps (Microsoft Visual Studio R and Intel Fortran compiler) should match with each other to avoid compatibility problems. A summary of the couples of versions that can be used can be found under Start > Programs > AspenTech > Process Modeling V.8.4. > Aspen Plus > Select compiler for Aspen Properties and Aspen Plus model libraries. Once Microsoft Visual Studio and Intel Fortran Compiler are installed, it is essential to proceed with the following steps:

1. Select Intel Fortran Compiler as the main compiler. To perform this operation, enter in the next route "start menu | All programs | AspenTech | Process Modeling V7.0| AspenPlus\Properties |Select compiler for Aspen Plus/Properties." A list of compilers will be displayed and select the correct option.
2. To compile the code, open the Aspen Plus Simulation Engine (see Figure 8.5).
3. Once the program is opened, the next windows, as shown in Figure 8.6, will be displayed.

By default an address appears that must be erased. Further it is essential to write the entire route where the file (FORTRAN code) is written, for example, "C:\Users\Gabriel\Desktop\Chapter." Ahead of this route, it is essential to write "aspcomp" followed by the name of the FORTRAN file. In this example, the complete line is aspcomp DNPECSB.f (see Figure 8.7).

As shown in Figure 8.6, if the compilation is correctly performed, two files with extension ".obj " and ".dll" with the same name of the code file will be automatically generated. Otherwise, the compilation errors must be

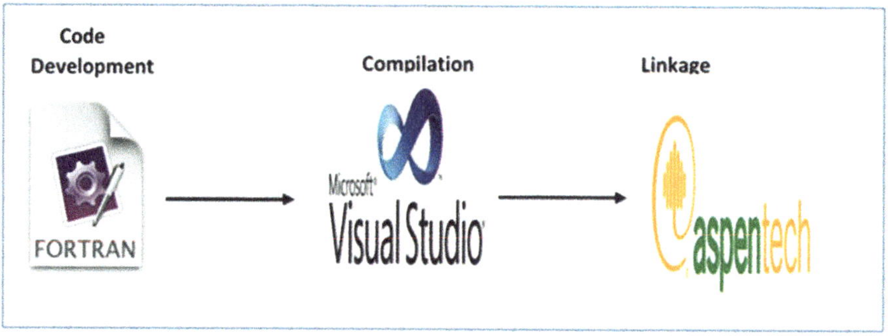

FIGURE 8.4
User Model for use in Aspen Plus.

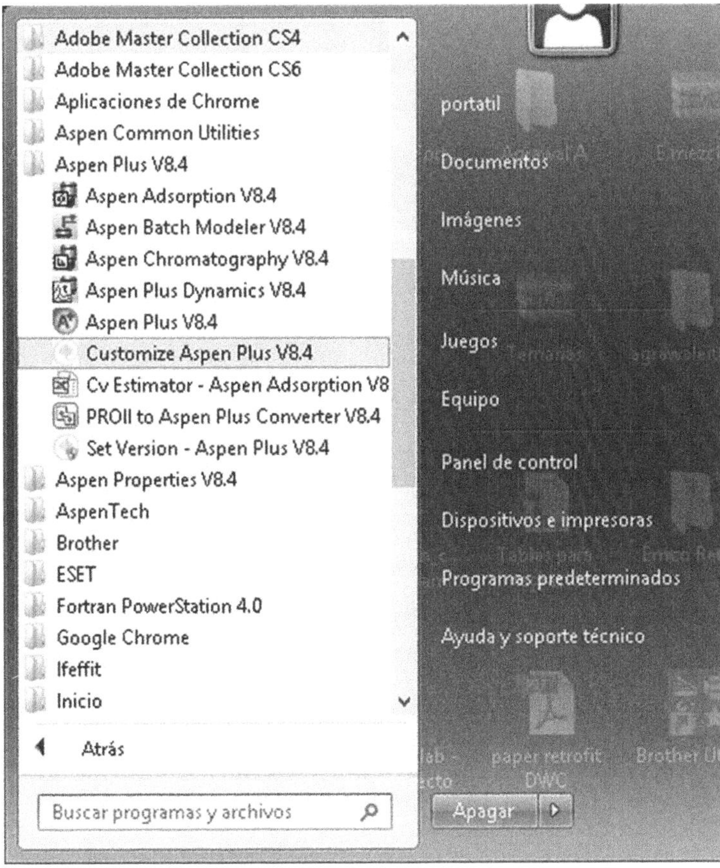

FIGURE 8.5
Opening Aspen Plus simulation engine.

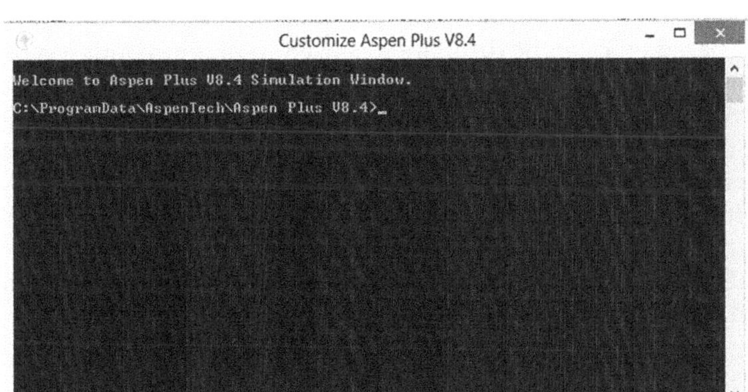

FIGURE 8.6
Aspen Plus simulation engine.

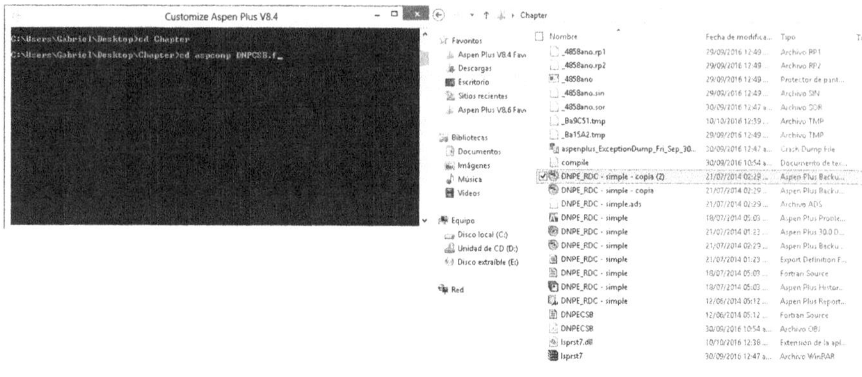

FIGURE 8.7
Results after compilation.

corrected. If these two files are created, this indicates that the link between Aspen Plus and FORTRAN is successfully established. Now, we are ready to work with Aspen Plus considering a FORTRAN Subroutine.

8.5 Optimization of a Reactive Distillation Column with a User Kinetic Model

As described in Section 8.1, the reactive distillation is an intensified unit operation, because of which the simulation of this process may turn complex and computational demanding. One example of the reactive distillation is the production of DNPE, which is an excellent candidate for diesel fuel formulations because of its blending cetane number. The process for DNPE production based on catalytic distillation is used to perform both reaction and separation in the same unit. The process flowsheet is schematically shown in Figure 8.8. The reactant is fed at the top of the reactive section, as saturated liquid stream. High-purity (over 99.9%) DNPE is obtained as bottom product, whereas the vapor distillate is condensed and sent to liquid–liquid separation, which provides the product water and the organic reflux (Dimian et al., 2014).

The dehydration of 1-pentanol to yield DNPE is an equilibrium limited reaction:

$$\underset{(P)}{2C_5H_9-OH} \leftrightarrow \underset{(DNPE)}{C_5H_9-O-C_5H_9} + \underset{(W)}{H_2O} \tag{8.1}$$

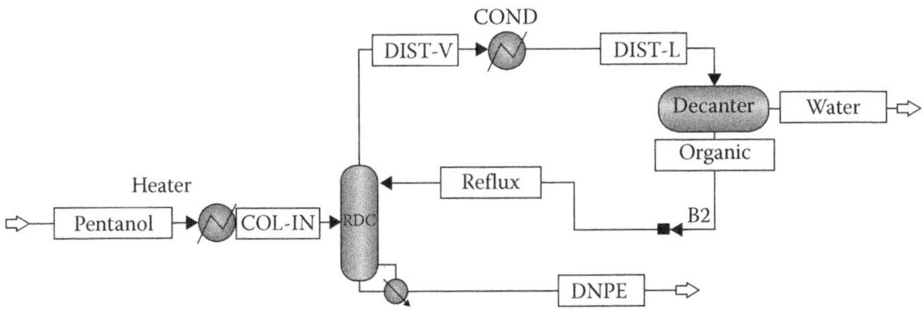

FIGURE 8.8
Catalytic distillation process for di-n-pentyl ether (DNPE) production.

The etherification of 1-pentanol is catalyzed by Amberlyst 70. Moreover, the reaction kinetics is described using the following expression, derived from the Eley–Rideal mechanism (Pera-Titus et al., 2009):

$$r = \frac{k a_P^2 \left(1 - \frac{1}{K_{eq}} \frac{a_W a_D}{a_P}\right)}{a_P \left(1 + K_W a_W^{1/2} a_P\right)} \tag{8.2}$$

where

$$k = 4.6 \times 10^{-6} \exp\left(-11{,}595\left(\frac{1}{T} - \frac{1}{438}\right)\right) \frac{\text{kmol}}{\text{kg}_{\text{cat}} \cdot \text{s}} \tag{8.3}$$

$$K_{eq} = 8.9229 \cdot \exp\left(\frac{778.69}{T}\right) \tag{8.4}$$

$$K_W = 4.306 \cdot \exp\left(-6616\left(\frac{1}{T} - \frac{1}{438}\right)\right) \tag{8.5}$$

To start the optimization process, it is necessary to prepare the Aspen File with the optimization program as described in Chapter 6. To start a simple design, it is necessary to introduce all components to be considered in the design (1-pentanol, water, and di-*n*-pentyl-ether), as shown in Figure 8.9. In this example, we have used Aspen Plus V8.4. Therefore, all the screen captures belong tho this version of the software.

The characterization of the feed stream is 25°C, 2 bar, and 42 kmol/h of 1-pentanol. This stream should be heated using a heater with the output conditions of 180°C, and 0 as vapor fraction. The module of distillation columns should be filled up with the conditions shown in Figures 8.10 through 8.13.

Once the module is ready, go to Reactions and select New. By default, this reaction will be named R-1. Moreover, in the type of reaction, select LHHW.

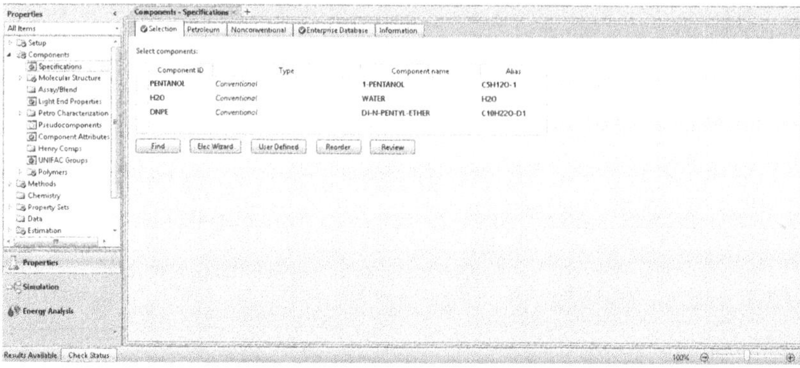

FIGURE 8.9
Introducing chemical components in Aspen Plus.

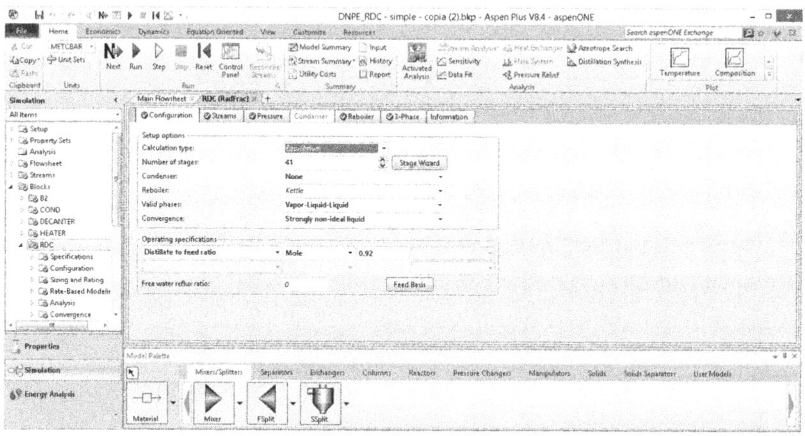

FIGURE 8.10
Partial input conditions of distillation module.

Then select the type of reaction as follows: kinetic, conversion, equilibrium, and click OK (see Figure 8.14).

In the following window, introduce the type of reaction, in this case Kinetic. In reactants, introduce 1-pentanol. The stoichiometric coefficient for 1-pentanol is 2 (automatically it will be changed as negative number), as shown in Figure 8.15. Next, introduce the products as follows: H_2O with coefficient 1 and DNPE with 1.

Furthermore, introducing the kinetic parameters in R-1, we select built-in Power Law and the reaction 1. Moreover, in Reacting phase, we select Liquid and introduce *k* value and the value of the activation energy in kcal/mol. Finally, we select the [C_i] basis in Cat(wt), as shown in Figure 8.16.

Furthermore, we introduce a second chemical reaction: in this case, a kinetic type. In reactants, we introduce 1-pentanol. The stoichiometric

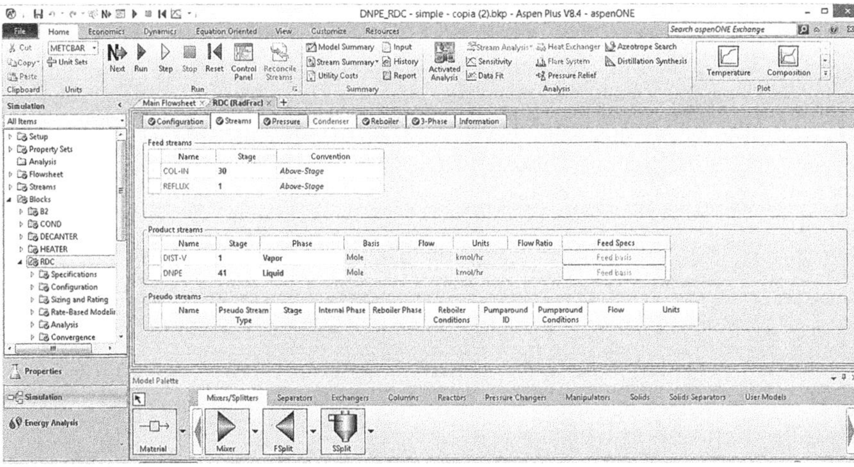

FIGURE 8.11
Stream input conditions of distillation module.

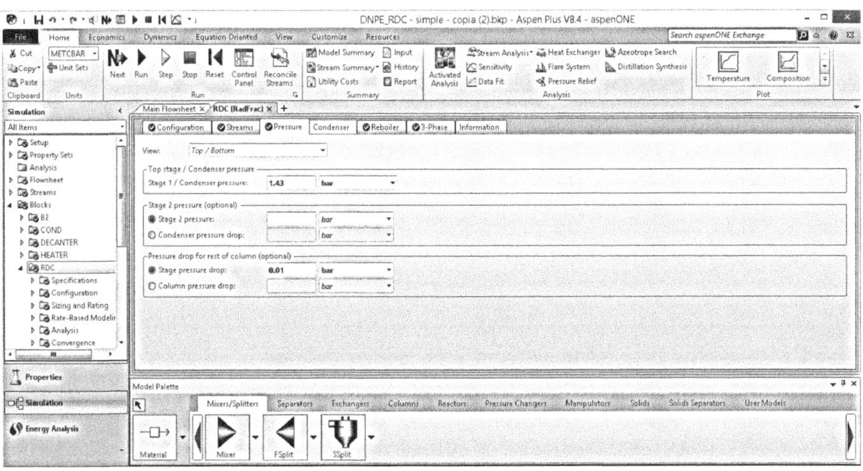

FIGURE 8.12
Pressure input conditions of distillation module.

coefficient for pentanol is 2 (stoichiometric coefficient is negative for reactive and positive for products). Next, we introduce the products as follows: H_2O with coefficient 1 and DNPE with 1.

However, this time, we select "user kinetic subroutine" and also "liquid" as reacting phase. In the section "subroutine," we introduce the name of the file (FORTRAN file) previously created and compiled (see Figure 8.17).

Now, we are ready to run this simulation by just clicking next in Aspen plus. Therefore, to perform the optimization process, it is necessary to develop the Excel Macro as described in Chapter 6. In this optimization

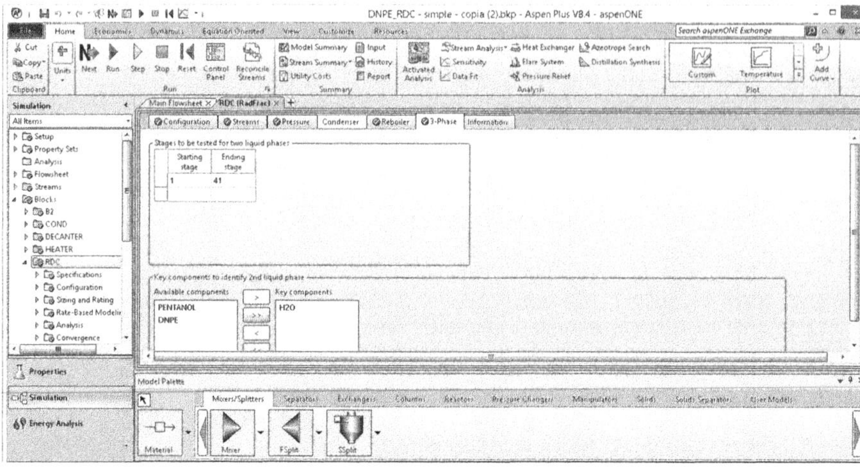

FIGURE 8.13
Three-phase input conditions of distillation module.

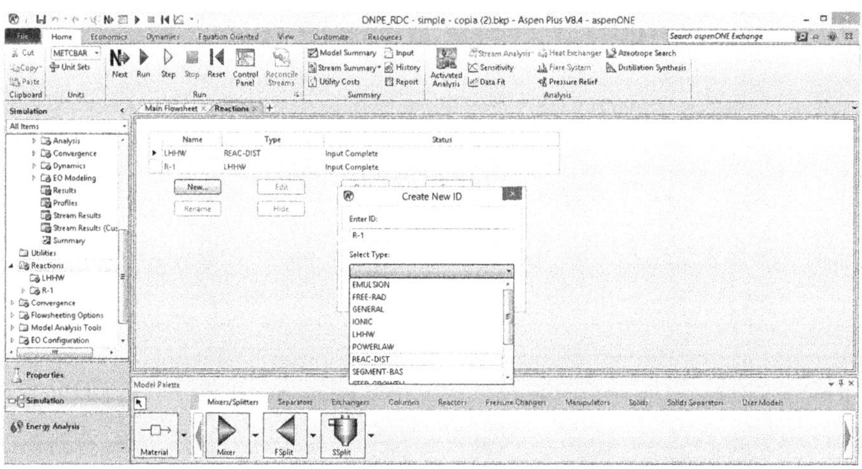

FIGURE 8.14
Selecting the type of reaction in Aspen Plus.

example, our objective function was the minimization of the total reboiler heat duty.

The minimization of this objective is subject to the required recoveries and purities in each product stream, i.e.,

$$\min(Q) = f\left(N_{tn}, N_{fn}, D_{rn}, R_{rn1}, R_{rn2}, P_{tn}, H_{tn}\right) \tag{8.6}$$

$$\text{Subject to } \vec{y}_m \geq \vec{x}_m$$

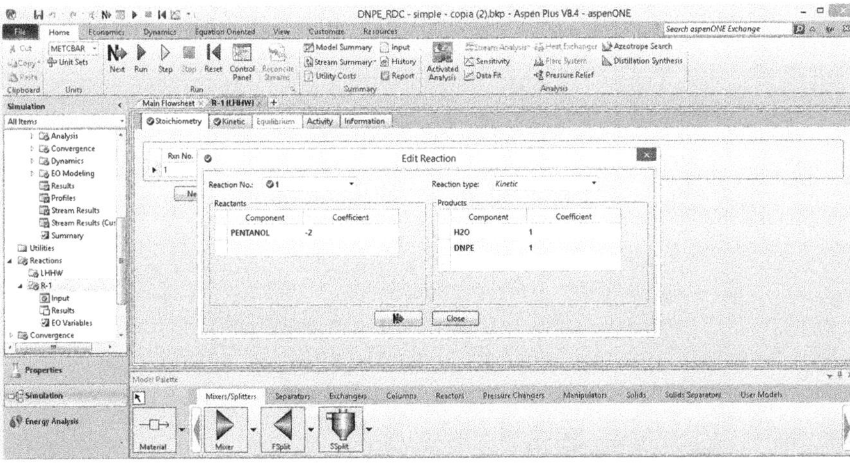

FIGURE 8.15
Filling up a chemical reaction in Aspen Plus.

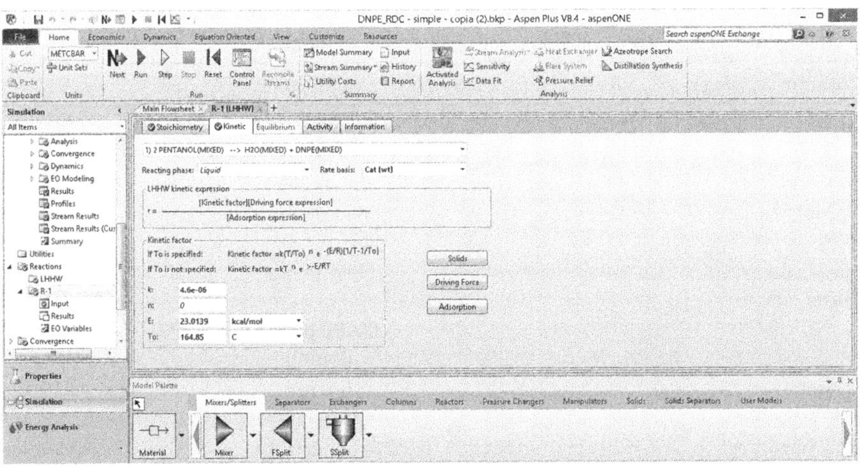

FIGURE 8.16
Kinetics parameters in R-1.

where N_{tn} are total column stages; N_{fn} is the feed stages in column; D_{rn} is the distillate to feed ratio D: $10 < D < 100$ kmol/h; R_{rn1} is the starting reactive stage; R_{rn2} is the ending reactive stage; P_{tn} is the operating pressure of the column P: $1 < P < 5$ bar; and H_{tn} is the holdup mcat: $20 < \text{mcat} < 200$ kg. The kinetics is in kmol/kg_cat/s, whereas y_m and x_m are vectors of the obtained and required purities for the m components, respectively. This minimization implies the manipulation of seven continuous and discrete variables. As later calculations, the total annual cost is also presented in Table 8.3.

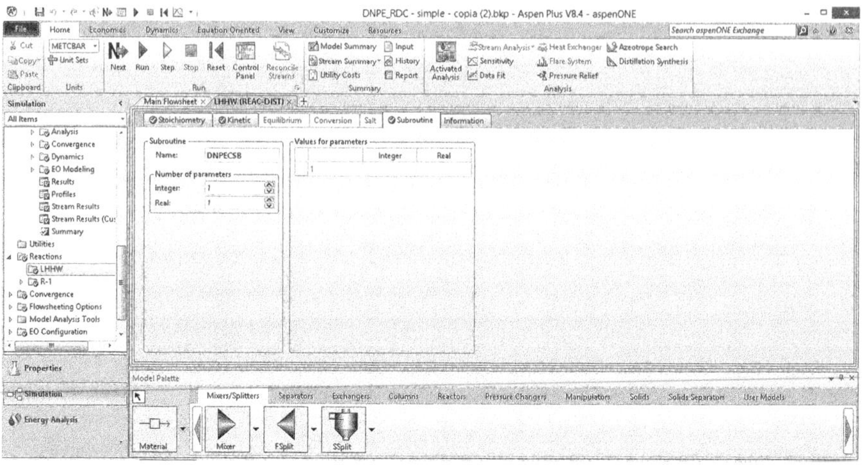

FIGURE 8.17
Calling a subroutine in Aspen Plus.

TABLE 8.3
Comparison of Reactive Distillation (RD) Process Alternatives

Column Topology	S1	S2	Specifications	S3	S4
Number of stages	41	–	Distillate to feed ratio	0.92	0.795
Feed stage	30	28	Boilup ratio	16.886	5.293
Starting reactive stage	11	4	Pressure (bar)	1.43	1.367
Ending reactive stage	40	28	Holdup (kg)	120	132.12
			DNPE purity (%mol)	0.9999	0.9999
Economic evaluation			**Energy requirements**		
Total operating cost ($/year)	2,453,162	888,084	Reboiler duty (Gcal/h)	3.415	1.057
Total investment cost ($/year)	1,14,813	123,326	Energy savings (%)	0	69.05
Total annual cost ($/year)	2,567,975	973,090			

After using the mono-objective optimization, it is possible to obtain a figure to observe the evolution of the objective function. This scenario is shown in Figure 8.18.

Figure 8.18 represents the evolution of objective function through the optimization process. Several points are highlighted: point S1 is the initial simulation, whereas the other simulations are produced in the optimization

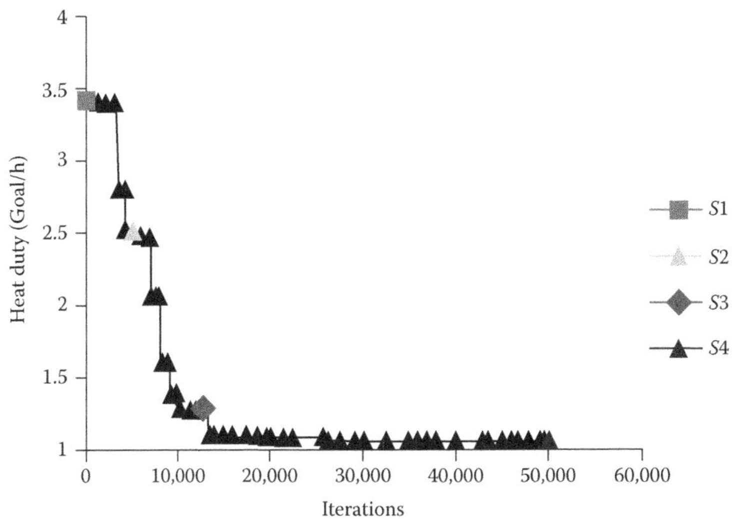

FIGURE 8.18
Evolution of objective function throughout optimization process.

process. Despite performing more function evaluations in our calculations, the optimization results are presented until 50,000 evaluations because the vector of decision variables does not produce a significant improvement. Under this scenario, it is assumed that differential evolution with tabu list (DETL) achieved the convergence at the tested numerical conditions, and the reported results correspond to the best solution obtained using the DETL method.

Table 8.3 shows the results of the initial and final points. Therefore, it is possible to observe that the reduction in energy consumption is approximately 69.0483%, which accomplishes purity and recovery constrains.

8.6 Reactive Distillation Column with a Default Kinetic Model

Inside Aspen Plus, it is also possible to perform a design similar to that presented in the first case of study in Section 8.5, without using a FORTRAN subroutine. In this section, this methodology is approached using a second case of study, the production of methyl *tert*-butyl ether (MTBE) (Calvo and Prieto, 2016). The MTBE is a gasoline component whose function is to increase the octane number in unleaded gasoline. The production and separation of this compound may be performed using a reactive distillation column. The problem to be solved is as laid out by Seader and Henley (2011): The reactive distillation column to produce MTBE is fed by one stream entering stage 10 consisting of 215.5 mol/s of methanol at 320 K and 11 bar, and another stream at stage 11 consisting of a mixture of 195.44 mol/s of isobutene and 353.56 mol/s of n-butene at 350 K and 11 bar. The column has 15 stages of liquid–vapor equilibrium, a total condenser, a partial reboiler, a reflux ratio equal to 7, and a bottom flow of 197 mol/s. The kinetic reaction between isobutene and methanol to produce MTBE according to Rehfinger and Hoffmann (1990) takes place in liquid phase, in a temperature range between 40°C and 100°C, using a strong-acid ion-exchange resin as catalyst (4.9 Eq/kg and 204.1 kg per stage). The column section where the reaction takes place is between stage 4 and stage 11. This design can be simulated in Aspen Plus, Aspen Plus V8.4 in this example, as illustrated in the following paragraphs.

First, it is necessary to introduce the components to be used: methanol, isobutene, 1-butene, and MTBE. To evaluate the thermodynamic interaction among all components, the Universal QuasiChemical-Redlich-Kwong (UNIQUAC-RK) model is considered. Therefore, the binary interaction parameters are presented in Figure 8.19.

The module to be used in RadFrac exhibits two separate feed streams: "Feed" and "Methanol." The stream "Feed" comprises 195.44 mol/s of isobutene and 353.56 mol/s of 1-butene at 350 K and 11 bar. The stream "Methanol" comprises 215.5 mol/s of methanol at 350 K and 11 bar.

The RadFrac Module must be filled up with the following parameters (see Figures 8.20 through 8.22):

- Calculation type: Equilibrium.
- Number of stages: 17.
- Condenser: Total.
- Convergence: Strongly nonideal liquid.

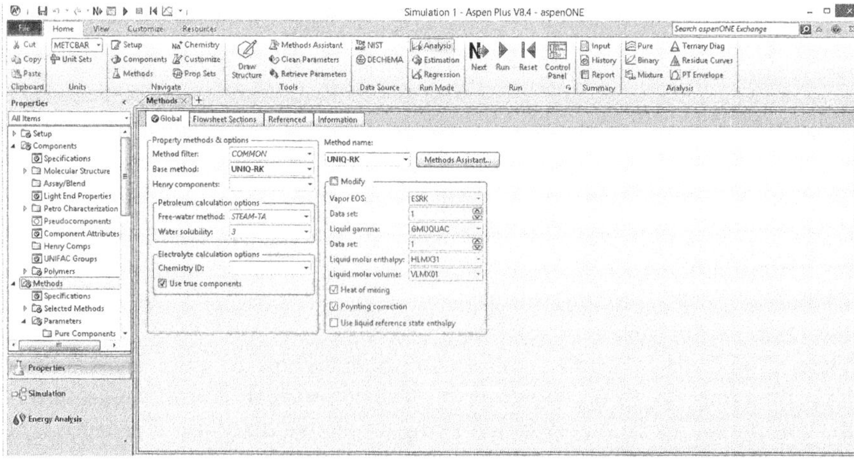

FIGURE 8.19
Binary interaction parameters in MTBE production.

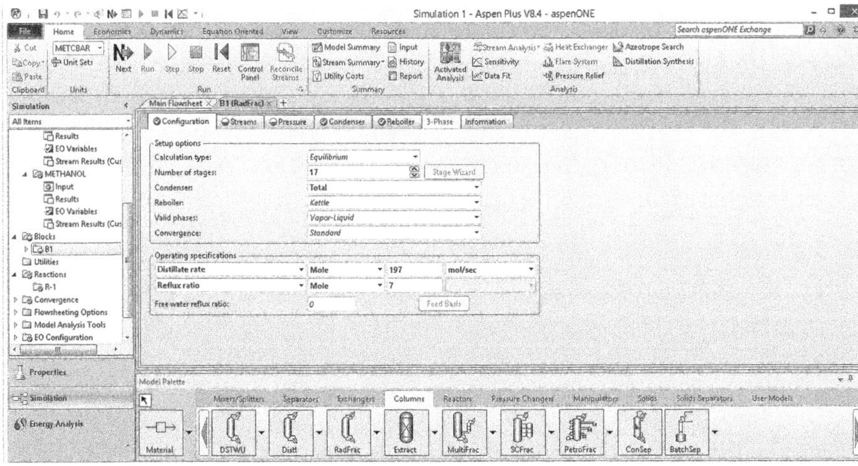

FIGURE 8.20
Input parameters of MTBE design.

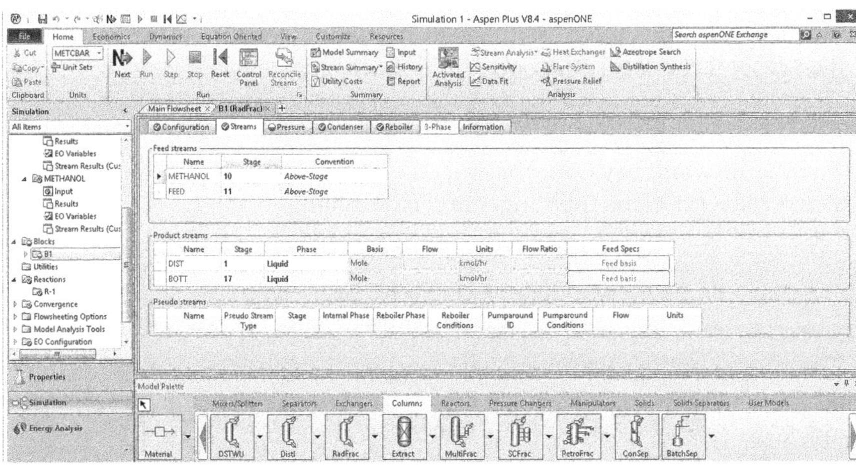

FIGURE 8.21
Feed streams in MTBE design.

- In Operating specifications, indicate that Bottom's rate is 197 mol/s and the Reflux ratio is 7.
- The stream "Feed" is located in stage 11, and the stream "Methanol" is located in stage 10.
- Condenser pressure: 11 bar.

To Add the entire reaction, we go to Reactions > New. We name it as R-1, and select the type of reaction: in this case REAC-DIST. We click on New and

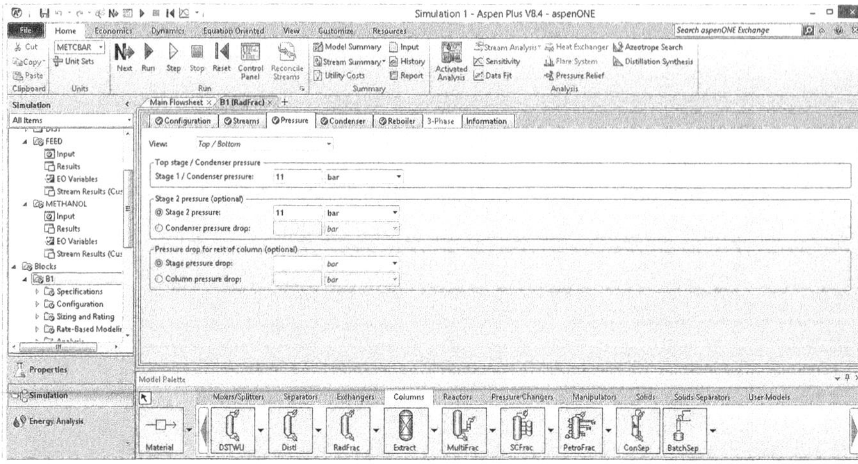

FIGURE 8.22
Pressure profile for input parameters in MTBE design.

Select the type of reaction as follows: kinetic, conversion, equilibrium, and then click OK.

The general chemical equation is as follows:

$$\text{IB} + \text{MeOH} \leftrightarrow \text{MTBE} \tag{8.7}$$

For the forward reaction, the rate law is formulated in terms of mole-fraction concentrations (Rehfinger and Hoffmann, 1990):

$$r_{\text{forward}} = 3.67 \times 10^{12} \cdot \exp\left(\frac{-92{,}440}{RT}\right)\frac{C_{IB}}{C_{MeOH}} \tag{8.8}$$

The corresponding backward rate law, consistent with chemical equilibrium, is as follows:

$$r_{\text{backward}} = 2.67 \times 10^{17} \cdot \exp\left(\frac{-134{,}454}{RT}\right)\frac{C_{MTBE}}{C_{MeOH}^2} \tag{8.9}$$

where r is in moles per second per equivalent of acid groups, R is 8.314 J/mol·K, T is in Kelvin, and C_i is the liquid mole fraction. Therefore, in the corresponding section, it is necessary to introduce the type of reaction, which in this case is kinetic. In reactants, we introduce isobutene and methanol. The stoichiometric coefficient for isobutene and methanol is 1 (which is negative for reactive and positive for products), and the exponent is 1 for the isobutene and –1 for the methanol. Next, we introduce the product as follows: MTBE with coefficient 1 and exponent 0. The same is followed for the backward reaction (see Figures 8.23 and 8.24).

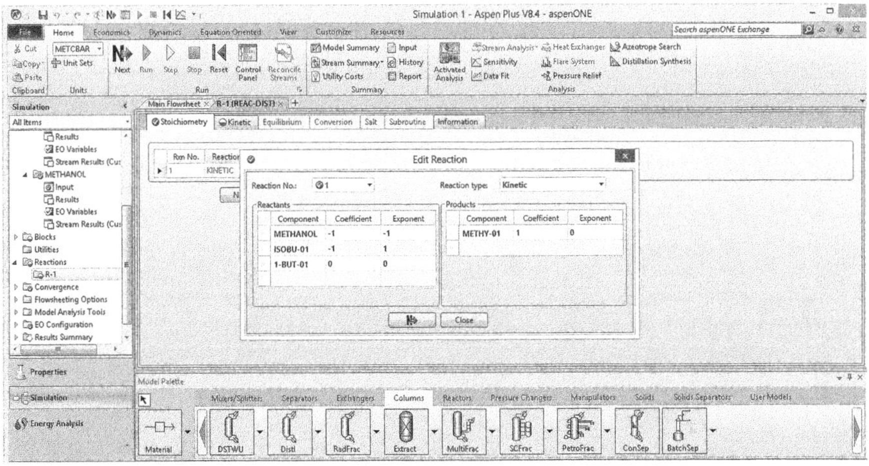

FIGURE 8.23
MTBE reaction in Aspen Plus.

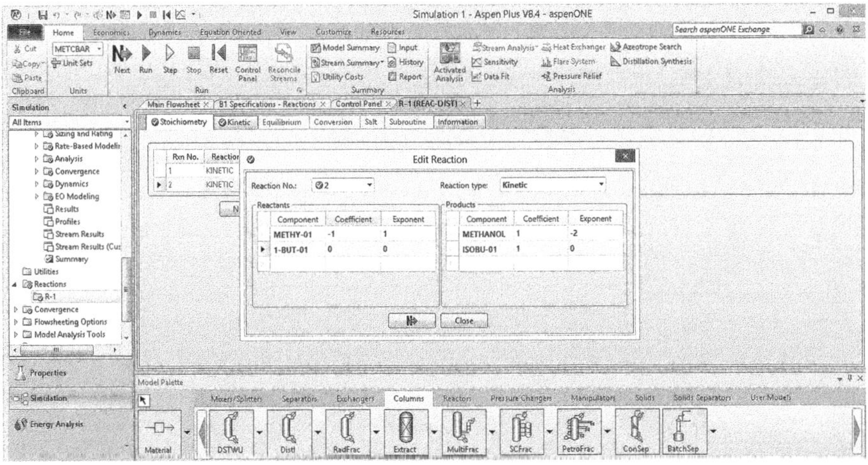

FIGURE 8.24
MTBE backward reaction in Aspen Plus.

Now, we introduce the parameters in the Kinetic tab. For each reaction, we select Use built-in Power Law, and select the reaction 1. In Reacting phase, we select Liquid, and introduce *k* value and the value of the activation energy in kJ/kmol. Finally, we select the [C_i] basis in Mole fraction, and the same is followed for the backward reaction (see Figures 8.25 through 8.27).

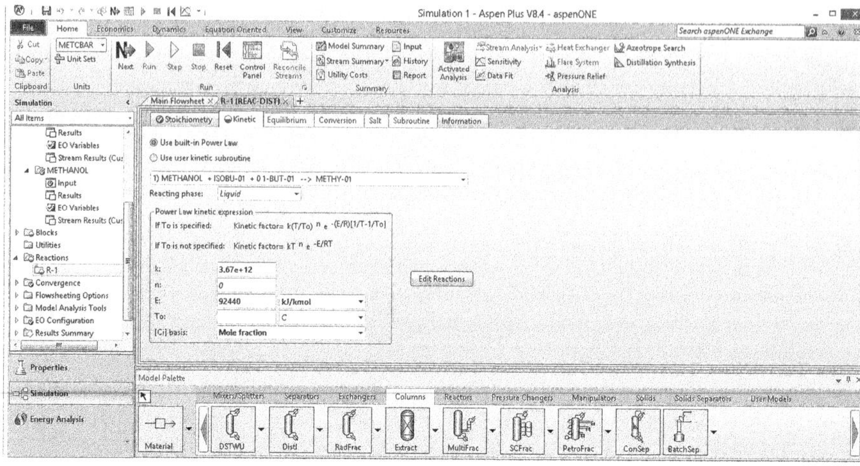

FIGURE 8.25
Kinetic parameters for MTBE reaction.

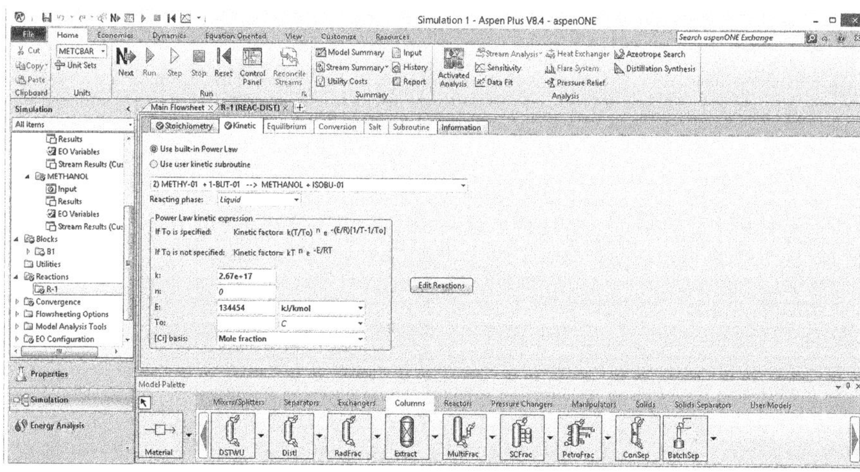

FIGURE 8.26
Kinetic parameters for backward MTBE reaction.

Now, in Blocks > C1 > Specifications > Reactions, we indicate that the reaction occurs in the column R-1, and between stage 4 and stage 11. Moreover, a holdup of 8000 kg occurs between stage 4 and stage 11, as shown in Figures 8.28 and 8.29. Finally, we are ready to run the complete Aspen Plus design and obtain the stream results and also a composition profile.

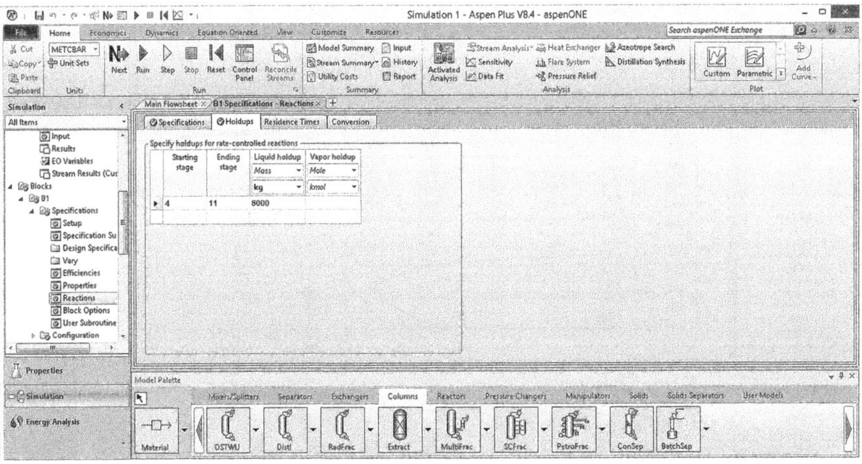

FIGURE 8.27
Reactive stages and holdup in MTBE reaction.

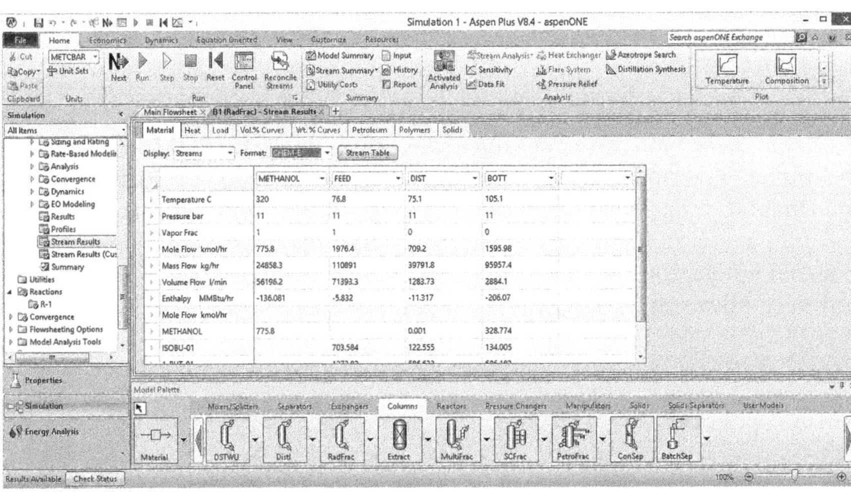

FIGURE 8.28
Streams results in MTBE reaction.

After these two cases of study, we are able to repeat this procedure for any type of process that involves a reaction. In this case, we use a distillation column to introduce a reaction inside the column; however, there are numerous possibilities.

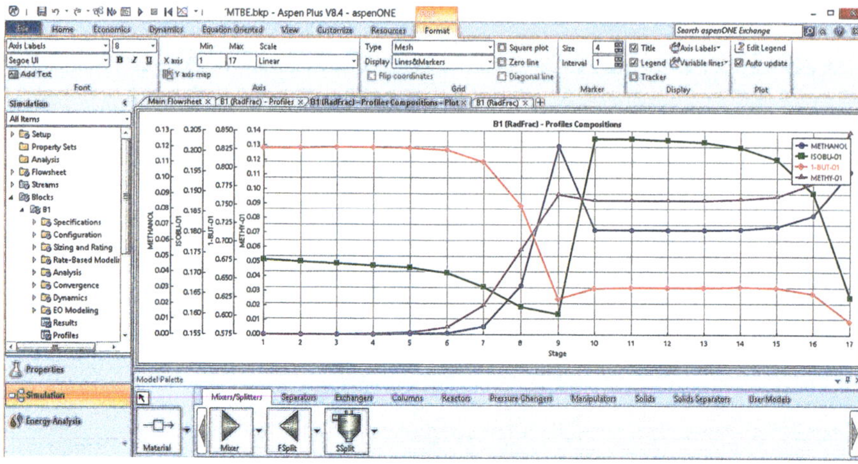

FIGURE 8.29
Composition profiles in MTBE reaction.

8.7 Conclusions

The process intensification is a very useful tool to improve the process; however, the current software does not always provide the necessary implements to simulate and analyze such intensified process. For the case discussed in this chapter, it has been necessary to write a model representing the kinetics of the reaction occuring in the reactive distillation system. The use of an external kinetic routine linked to Aspen Plus extendes the capabilities of the software, allowing the user the simulation of non-conventional kinetics.

References

Aspen Technology Inc. (2001), Aspen Plus 11.1 User Guide, Cambridge, MA, U.S.A.

C.S. Bildea, R. György, E. Sánchez-Ramírez, J.J. Quiroz-Ramírez, J.G. Segovia-Hernandez, and A.A. Kiss, 2015, Optimal design and plantwide control of novel processes for di-*n*-pentyl ether production. *J. Chem. Technol. Biotech.*, 90(6), 992–1001.

A.C. Dimian, C.S. Bildea, and A.A. Kiss, Integrated design and simulation of chemical processes, 2nd edition, Elsevier, Amsterdam, 2014.

M.F. Doherty and G. Buzzad, 1992, Reactive distillation by design. *Trans. Icehm*, 70, 448–458.

M. Pera-Titus, J. Llorens, and F. Cunill, 2009, Technical and economical feasibility of zeolite NaA membrane-based reactors in liquid-phase etherification reactions. *Chem. Eng. Process.*, 48,1072–1079.

A. Rehfinger and U. Hoffmann, 1990, Kinetics of methyl tertiary butyl ether liquid phase synthesis catalyzed by ion exchange resin—I. Intrinsic rate expression in liquid phase activities. *Chem. Eng. Sci.*, 45(6), 1605–1617.

J.D. Seader and E.J. Henley, 2011, Separation process principles. Hoboken, NJ: Wiley.

R. Turton, R.C. Bailie, W.B. Whiting, J.A. Shaeiwitz, 2009, *Analysis, Synthesis and Design of Chemical Processes*, 3rd edition, Upper Saddle River, NJ: Prentice Hall (Appendix A).

G.D. Ulrich, 1984, *A Guide to Chemical Engineering Process Design and Economics*, New York, NY: Wiley.

E. Vlad, C.S. Bildea, G. Bozga, 2013, Robust optimal design of a glycerol etherification process, *Chem. Eng. Technol.*, 36, 251–258.

9

*Optimization of a Biobutanol Production Process**

9.1 Introduction

Recently, renewable energy sources have been considered sustainable because they apparently do not affect the global food production in terms of competing energy and food use (Rathmann et al., 2010). Considering the life-cycle assessment of second- or third-generation biofuel feedstocks, such as crops, forest residue, wood process waste, organic portion of municipal waste, or algae, it has been found that these feedstocks represent a vastly underutilized renewable power source. Biofuel production from such feedstocks would drastically reduce greenhouse gas emissions, thereby providing a sustainable future.

Currently, there is global demand for butanol and it is produced by petrochemical methods (Rathmann et al., 2010). Acetone–butanol–ethanol (ABE) fermentation is a commercial process to produce butanol, acetone, and ethanol using microorganisms. Several solventogenic *Clostridium* species are used in this process (Sorda et al., 2010). To design a fermenter to produce butanol, it is necessary to predict the metabolic behavior according to the needs of the process. The classic metabolic route of solventogenic *Clostridium* species is shown in Figure 9.1.

The biobutanol can be produced by several strains and these cultures include *Clostridium acetobutylicum* P262, *C. acetobutylicum* ATCC 824, *C. acetobutylicum* NRRL B643, *C. acetobutylicum* B18, *C. beijerinckii* P260, *C. beijerinckii* BA101, *C. beijerinckii* LMD 27.6, *C. butylicum*, *C. aurantibutyricum*, and *C. tetanomorphum*. Among all these strains, *C. saccharobutylicum* P262 (formerly known as *C. acetobutylicum* P262) and *C. beijerinckii* P260 were used for commercial purpose in South Africa until the early 1980s (Zverlov et al., 2006). Currently, almost all research efforts are focused on genetically improving butanol-producing cultures. These cultures can produce or tolerate elevated levels of ABE fluctuating from 20 to 30 g/L (Jones and Woods, 1986), which explains the increase in the number of studies on this aspect in recent years. Despite the inhibition effects in fermentations, significant improvements in the tolerance level of butanol have been reported.

* Eduardo Sánchez-Ramírez and Juan José Quiroz-Ramírez also contributed to the work on this chapter.

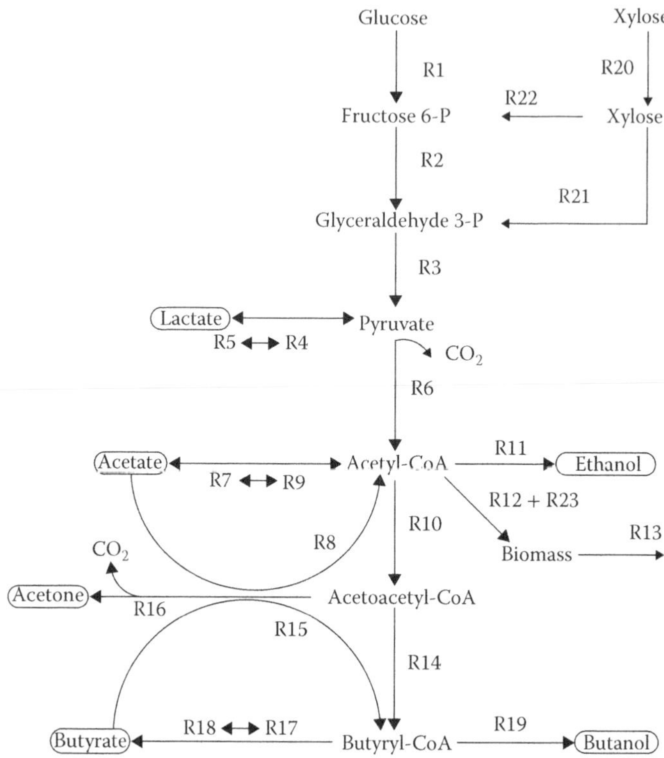

FIGURE 9.1
Metabolic pathway for xylose and glucose consumption.

There are several studies on the metabolism of solventogenic *Clostridium* species, which comprises two clearly distinctive phases: solventogenics and acidogenics (Li et al., 2011). In the acidogenic phase, some acids are produced in the exponential growth phase. Further in the solventogenic phase, these acids are assimilated and all other components of interest (acetone, butanol, and ethanol) are produced. This fermentation process is quite complex, however this is not a hurdle to develop a mathematic model to describe it (Shinto et al., 2007, 2008; Kim et al., 2012; Junne, 2010). Shinto et al. (2007, 2008) proposed an adequate model to describe the glucose and xylose consumed by microorganisms to produce acetone, butanol, and ethanol.

Important limitations of ABE fermentation include low yield and high production cost. Lignocellulosic biomass offers important properties that help diminish such limitations. This lignocellulosic biomass needs to be pretreated; in other words, it is necessary to reduce hemicellulose to xylose and also to reduce the crystallinity of cellulose (Jang et al., 2012). After pretreatment, lignocellulosic biomass should be hydrolyzed either chemically or enzymatically (Chen and Wang, 2010).

Enzymatic hydrolysis offers several advantages over both physical and chemical mechanisms; for example, it reduces cost by diminishing energy requirements. Despite enzymatic hydrolysis is highly inhibited by the products glucose and xylose, the use of an integrated reactor (on which fermentation and saccharification occur simultaneously) decrease such inhibition, since all monosacharides are consumed in the fermentation. (Andrić et al., 2010).

Integration of hydrolysis, fermentation, and recovery in ABE process was suggested by Qureshi et al. (2014). In simultaneous saccharification and fermentation (SSF), hydrolysis and fermentation are carried out in a single operation unit. As a result of reduction in sugar retention inside the reactor, the SSF process increases yields and decreases energy requirements.

It is known that ABE fermentation has low yields; at a concentration of 15 g/L, the fermentation process is significantly inhibited. Under this scenario, integrated reactors with a separation unit have proved to reduced butanol inhibition. Several techniques have been proposed for the separation unit, highlighting adsorption, liquid–liquid extraction, pervaporation, and gas stripping (Qureshi and Blaschek, 2001; Abdehagh et al., 2014).

With this in mind, the butanol production by ABE fermentation is a process that needs to be optimized to minimize all costs and energy requirements. However, the entire kinetics must be correctly modeled to predict the fermentation process. Furthermore, all parameters in the separation unit need to be determined to correctly predict the equilibrium among all components. In this chapter, we focus on the optimization of butanol production to obtain designs that accomplish economic and environmental targets. We consider liquid–liquid extraction as a unit separation and an integrated reactor to perform the entire fermentation.

9.2 Description of the Process

Biobutanol can be produced by fermentation using *Clostridium* bacteria. The simplified process is shown in Figures 9.2 and 9.3. This study is based on the report presented by Díaz and Tost (2016). In this case, a reactor of 1000 m^3 is considered with one feed with constant sugar concentration. Note that a purge has been used to control the volume of fermentation, which helps removing all these components without reaction and, consequently, also reduces the inhibition factor.

In this work, simultaneous fermentation, saccharification, and extraction are rigorous modeled in MATLAB®. Shinto et al. (2007, 2008) developed a rigorous mathematical model with multiple metabolic reactions to describe the dynamic profile considering all intermediate components involved in fermentation, and considering all inhibitory effects. A key concept of this model is the inclusion of glucose and xylose in the metabolic route of ABE fermentation, thus this model should be adjusted to experimental data. On the other

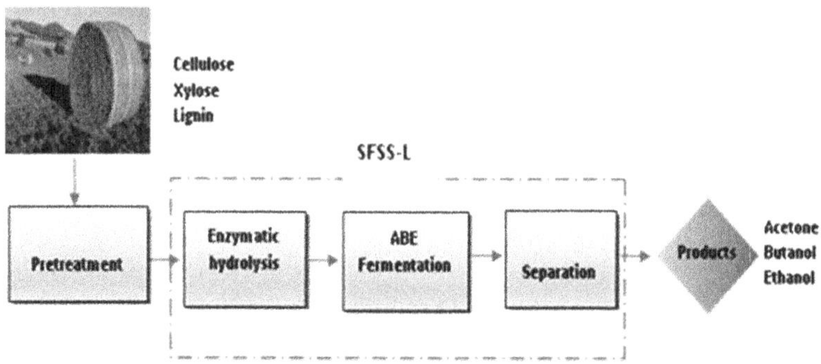

FIGURE 9.2
Two flowsheet processes of biobutanol production.

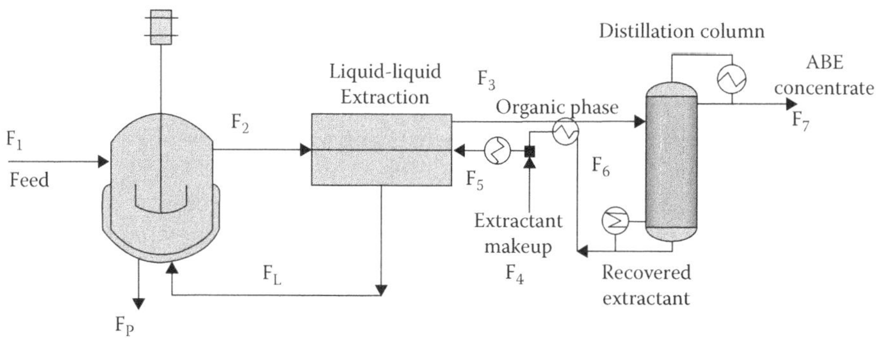

FIGURE 9.3
Complete flowsheet of the case of study in this work.

hand, the saccharification model is based on the work of Kadam et al. (2004). Note that in this work, we fix cellulose/xylose/lignin ratio of 2:1.33:1 (g/L).

In this study, we assumed the solids as empty fraction in volume calculation, which implies a lost in the volume available for reaction (V_F). The global kinetic saccharification reaction (R) has been obtained by the ratio of the kinetic model of Kadam et al. (2004) and the flow rate (V_F), which allows us to turn this variable as continuous in all fermentation processes. The purge flows (F_p) have been obtained by implementing a proportional controller that monitors the total volume (V_T). Thus, the general balance of the integrated reactor is stated as follows:

$$\frac{dC_i}{dt} = R_i \cdot V_F + F_1 \cdot x_{Li} - F_p \cdot x_{Pi} + F_L \cdot x_{Li} \cdots \quad (9.1)$$

where C_i is the concentration of each component (butanol, ethanol, acetone, butyric acid, acetic acid, glucose, and xylose), x_{Li} is the composition of the component i in the stream F_L, F_L is the stream returning to the reactor, x_{Pi} is the composition of the component i in the stream F_p.

9.3 Thermodynamics and Kinetic Model

After obtaining the model of the ABE process, the next step is to obtain the thermodynamic parameters of all equations. In this work, we have considered *Clostridium saccharoperbutylacetonicum* N1-4, and xylose and glucose as substrates. Table 9.1 shows the initial condition for kinetic calculation.

Shinto et al. (2008) proposed an "on–off" method to describe the metabolism of ABE fermentation according to experimental data. In this study, we adjusted experimental data from eight batch fermentations, four from glucose and four from xylose, at a pH of 6.5 and a temperature of 30°C.

The adjustment of experimental data into Shinto model was made using optimization toolbox in MATLAB considering a sum of 58 parameters. In this optimization process, Microsoft Excel is necessary for exchange of data (Figure 9.4).

In this optimization process, the objective function to be accomplished is the difference between the norm of experimental data and simulation data divided by its maximum experimental data (see Equation 9.2):

$$F_{\text{ob}} = \sqrt{\sum_{i=1}^{n}\sum_{j=1}^{m}\left(\frac{c_{eij} - c_{sij}}{c_{j\max}}\right)^2} \tag{9.2}$$

Thus, the objective function is written as follows:

$$\begin{gathered} \text{Min}(F_{\text{obj}}) = \text{f}(V_{\max j}, K_{mi}, K_{maj}, K_{mbj}) \\ \textit{subject to:} \\ \vec{y}_m \geq \vec{x}_m \end{gathered} \tag{9.3}$$

TABLE 9.1

Initial Conditions of the Fermentations To Be Studied

Fermentation	Glucose (g/L)	Xylose (g/L)	Acetate (g/L)	Butyrate (g/L)	Butanol (g/L)	Acetone (g/L)	Biomass (g/L)
F_1	53.1	–	2.47	0.09	0.33	0.17	0.24
F_2	22.0	–	2.47	0.09	0.21	0.02	0.24
F_3	12.7	–	2.79	0.38	0.30	0.12	0.22
F_4	6.5	–	2.50	0.17	0.22	0.02	0.22
F_5	–	43.8	2.42	0.23	0.16	0.05	0.14
F_6	–	17.4	2.45	0.06	0.15	0.05	0.14
F_7	–	9.9	1.80	0.10	0.15	0.09	0.14
F_8	–	6.1	2.27	0.10	0.16	0.08	0.14

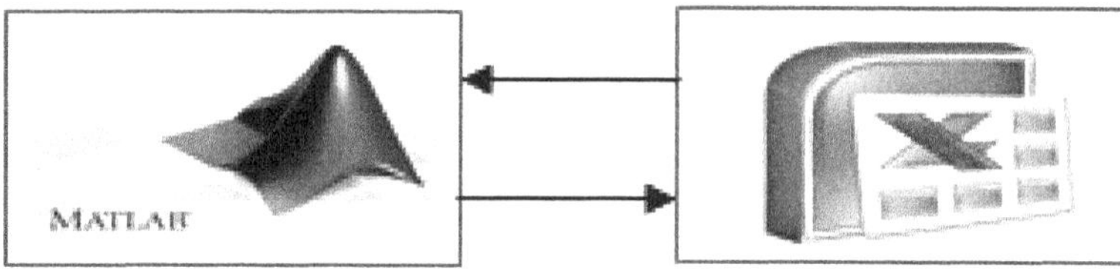

FIGURE 9.4
Optimization platform.

where V_{maxj} is the maximum reaction rate in reaction j, K_{mi} is the Michaelis–Menten constant (mM) in reaction j, K_{maj} is the activation constant (mM) for activators in reaction j, K_{mbj} is the inhibition constant (mM) for inhibitors in reaction j, y_m and x_m are vectors of obtained and required value for the m parameter, respectively.

The kinetic model used in this work is described by Equations 9.4 through 9.25, r (mmol/h):

$$r_1 = \frac{V_{max1}[\text{Glucose}][\text{Biomass}]}{K_{m1}\left(1+\frac{[\text{Glucose}]}{K_{m1A}}\right)+[\text{Glucose}](1+[\text{Butanol}])}.F \tag{9.4}$$

$$r_2 = \frac{V_{max2}[FGP][Biomass]}{K_{m2}+[FGP]}.F \tag{9.5}$$

$$r_3 = \frac{V_{max3}[\text{G3P}][\text{Biomass}]}{K_{m3}+[\text{G3P}]}.F \tag{9.6}$$

$$r_4 = \frac{V_{max4}[\text{Lactate}][\text{Biomass}]}{K_{m4}+[\text{Lactate}]}.F \tag{9.7}$$

$$r_5 = \frac{V_{max5}[\text{Pyruvate}][\text{Biomass}]}{K_{m5}+[[\text{Pyruvate}]]}.F \tag{9.8}$$

$$r_6 = \frac{V_{max6}[[\text{Pyruvate}]][\text{Biomass}]}{K_{m6}+[\text{Pyruvate}]}.F \tag{9.9}$$

$$r_7 = \frac{V_{max7}[\text{Acetate}][\text{Biomass}]}{K_{m6}+[\text{Acetate}]}.F \tag{9.10}$$

$$r_8 = V_{\max 8}\left(\frac{1}{1+[K_{m8A}/\text{Acetate}]}\right)\left(\frac{1}{1+[K_{m8B}/\text{Acetate}]}\right)[\text{Biomass}] \tag{9.11}$$

$$r_9 = \frac{V_{\max 11}[\text{ACoA}][\text{Biomass}]}{K_{m11}+[\text{ACoA}]}\cdot F \tag{9.12}$$

$$r_{10} = \frac{V_{\max 11}[\text{ACoA}][\text{Biomass}]}{K_{m11}+[\text{ACoA}]}\cdot F \tag{9.13}$$

$$r_{11} = \frac{V_{\max 11}[\text{ACoA}][\text{Biomass}]}{K_{m11}+[\text{ACoA}]}\cdot F \tag{9.14}$$

$$r_{12} = \left(\frac{V_{\max 12}[\text{Glucose}][\text{Biomass}]}{\text{Glucose}+K_{m12}+K_{m12A}/[\text{ACoA}]}\right).\text{INH} \tag{9.15}$$

$$r_{13} = k_{13}[\text{Biomass}]$$

$$r_{14} = \frac{V_{\max 18}[\text{AACoA}][\text{Biomass}]}{K_{m18}+[\text{AACoA}]}\cdot F \tag{9.16}$$

$$r_{14} = \frac{V_{\max 18}[\text{AACoA}][\text{Biomass}]}{K_{m18}+[\text{AACoA}]}\cdot F \tag{9.17}$$

$$r_{15} = V_{\max 15}\left(\frac{1}{1+(K_{m15A}/[\text{Butyrate}])}\right)\left(\frac{1}{1+(K_{m15B}/[\text{AACoA}])}\right)[\text{Biomass}] \tag{9.18}$$

$$r_{16} = r_8 + r_{15}$$

$$r_{17} = \frac{V_{\max 17}[\text{Glucose}][\text{Biomass}]}{K_{m17}+[\text{Butyrate}]+K_{m17A}/[\text{Butyrate}]+K_{m17B}/[\text{Butanol}]}\cdot F \tag{9.19}$$

$$r_{18} = \frac{V_{\max 18}[\text{BCoA}][\text{Biomass}]}{K_{m18}+[\text{BCoA}]}$$

$$r_{19} = \frac{V_{\max 19}[\text{BCoA}][\text{Biomass}]}{K_{m19}(1+K_{m19A}/[\text{Butyrate}])+[\text{BCoA}](1+[\text{Butanol}]/K_{m19B})}\cdot F \tag{9.20}$$

$$r_{20} = \frac{V_{\max 20}[\text{Xylose}][\text{Biomass}]}{K_{m20}\left(1+[\text{Xylose}]/K_{m20\text{A}}\right)+[\text{Xylose}]\left(1+[\text{Butanol}]\right)} \cdot F \quad (9.21)$$

$$r_{21} = \frac{V_{\max 21}[\text{X5P}][\text{Biomass}]}{K_{m21}+[\text{X5P}]} \cdot F \quad (9.22)$$

$$r_{22} = \frac{V_{\max 22}[\text{X5P}][\text{Biomass}]}{K_{m22}+[\text{X5P}]} \cdot F \quad (9.23)$$

$$r_{23} = \left(\frac{V_{\max 23}[\text{Xylose}][\text{Biomass}]}{[\text{Xylose}]+K_{m23}+K_{m23\text{A}}/[\text{ACoA}]}\right) \cdot \text{INH} \quad (9.24)$$

$$\text{INH} = \left(1-\frac{\text{Butanol}}{\text{Bu}_{\max}}\right)\cdot\left(1-\frac{\text{Acetate}}{\text{A}_{\max}}\right)\cdot\left(1-\frac{\text{Butyrate}}{\text{But}_{\max}}\right)\cdot\left(1-\frac{\text{Biototal}}{\text{Bio}_{\max}}\right) \quad (9.25)$$

where F is equal to 1 if substrate concentration is higher than 1 (mM). On the contrary, it is equal to zero and this is why this is called the "on–off" model. To solve the entire model, we also considered Equations 9.26 through 9.40 to describe the rate of all involved components during fermentation:

$$\frac{d[\text{Glucose}]}{dt} = -r_1 \quad (9.26)$$

$$\frac{d[\text{F6P}]}{dt} = r_1 - r_2 + r_{21} \quad (9.27)$$

$$\frac{d[\text{Pyruvate}]}{dt} = r_3 + r_4 + r_5 - r_6 \quad (9.28)$$

$$\frac{d[\text{Lactate}]}{dt} = r_5 - r_4 \quad (9.29)$$

$$\frac{d[\text{ACoA}]}{dt} = r_6 + r_7 + r_8 - r_9 - r_{10} - r_{11} - r_{12} - r_{23} \quad (9.30)$$

$$\frac{d[\text{Biomass}]}{dt} = r_{12} + r_{23} - r_{13} \quad (9.31)$$

$$\frac{d[\text{Acetate}]}{dt} = r_9 - r_7 - r_8 \tag{9.32}$$

$$\frac{d[\text{Ethanol}]}{dt} = r_{11} \tag{9.33}$$

$$\frac{d[\text{BCoA}]}{dt} = r_{14} + r_{15} + r_{17} - r_{18} - r_{19} \tag{9.34}$$

$$\frac{d[\text{Butyrate}]}{dt} = r_{18} - r_{15} - r_{17} \tag{9.35}$$

$$\frac{d[\text{Acetone}]}{dt} = r_{16} \tag{9.36}$$

$$\frac{d[\text{Butanol}]}{dt} = r_{19} \tag{9.37}$$

$$\frac{d[\text{CO2}]}{dt} = r_{16} + r_6 \tag{9.38}$$

$$\frac{d[\text{Xylose}]}{dt} = -r_{20} \tag{9.39}$$

$$\frac{d[\text{X5P}]}{dt} = -r_{22} - r_{21} + r_{20} \tag{9.40}$$

Once the optimization process is concluded, we obtain all parameters. Table 9.2 shows these parameters for biobutanol. Note further these parameters are adjusted to the experimental data reported, so that we can predict the concentration of any specie in fermentation.

A comparison between the predictions of the kinetic model and the experimental data for the fermentation of glucose and xilose, can be observed in Figure 9.5.

The predicted behavior through the optimization process is acceptable for glucose, biomass, butanol, acetone, and ethanol, which represents products of our interest. In this example, the parameters for the kinetic models are shown in Table 9.3. In this manner, it is possible to implement an optimization process to adjust kinetic parameters in any model.

On the other hand, note we are proposing a separation unit at the end of the integrated reactor. However, it is also necessary both to correctly select

TABLE 9.2
Adjusted Parameters for the Proposed Model

Species	F_1	F_2	F_3	F_4	F_5	F_6	F_7	F_8	Average
Butanol	0.985	0.967	0.964	0,991	0.994	0.960	0.986	0.963	0.976
Glucose	0.993	0.976	0.968	0,985	–	–	–	–	0.980
Acetone	0.954	0.994	0.993	0,954	0.973	0.963	0.918	0.930	0.960
Acetate	0.873	0.990	0.982	0,909	0.837	0.778	0.834	0.885	0.886
Biomass	0.740	0.842	0.861	0,988	0.866	0.900	0.915	0.974	0.886
Butyrate	0.822	0.778	0.950	0,893	0.936	0.862	0.838	0.830	0.864
Xylose	–	–	–	–	0.997	0.979	0.977	0.995	0.987
Average	0.895	0.925	0.953	0.953	0.934	0.907	0.911	0.929	0.926

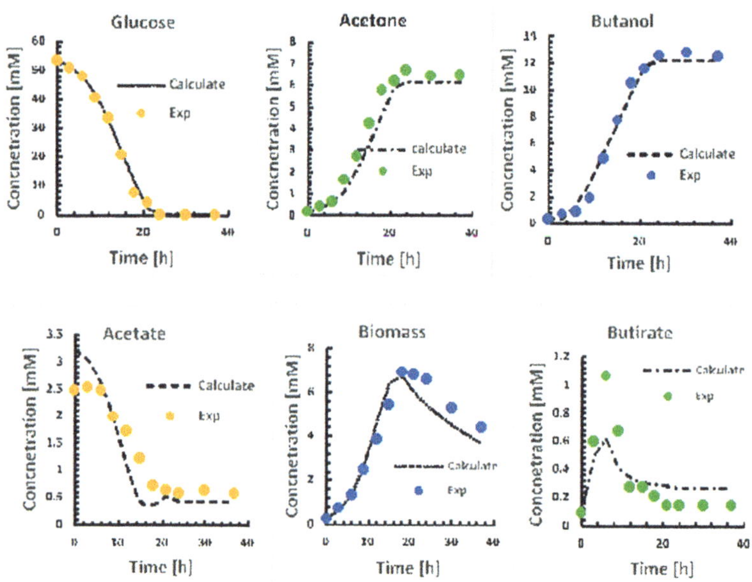

FIGURE 9.5
Prediction of acetone–butanol–ethanol (ABE) species in fermentation with kinetic parameters.

the extractant and to predict the equilibrium among ABE mixture and the extractant. Adding the extractant allows us to break both azeotropes in the ABE mixture, a homogeneous mixture between ethanol and water and a heterogeneous mixture between butanol and water. In this example, we decided to use 2-ethyl-1-hexanol because of its high selectivity, high partition coefficient (8), low cost (0.93$/kg), and high boiling temperature (185°C) (González-Peñas et al., 2014).

TABLE 9.3

Adjusted Parameters for the Proposed Model

Reaction	V_{max}	K_m	K_{ma}	K_{mb}
r_1	1.90	5.45	93.11	39.49
r_2	21.89	1.54	–	–
r_3	42.15	4.09	–	–
r_4	2.15	173.16	–	–
r_5	7.47	131.35	–	–
r_6	67.64	3.12	–	–
r_7	19.50	22.21	–	–
r_8	8.08	4.59	45.04	
r_9	34.84	0.71	–	–
r_{10}	33.98	10.99	–	–
r_{11}	4.60	7.40		–
r_{12}	0.53	18.25	6.37	–
r_{13}	0.019	–	–	–
r_{14}	92.47	16.46	–	–
r_{15}	0.23	2.20	0.00105	–
r_{16}	11.96	0.48	24.74	175.29
r_{17}	31.10	1.33		
r_{18}	92.81	1.30	86.04	148.15
r_{19}	1.55	3.33	51.80	96.20
r_{20}	99.47	4.66	–	–
r_{21}	131.35	25.78	–	–
r_{22}	0.15	0.06	0.11	–
r_{23}	1.90	5.45	93.11	39.49

```
%thermodynamic parameters are computed for the UNIQUAC model
%
clc, clear all, close all
% 1 Butanol
% 2 Ethanol
% 3 Acetone
% 4 Water
% 5 2-etheyl-1-hexanol
%Initial parameters used to start the iterations
Para=[-162.993914762840 838.413926021204 100.818705473944
-195.901793585449 50.1756052779197 -174.669148161186
252.772011901856 -723.591282305885 -525.317586936387
220.675537041103 0.0759244134280697 -4.02005739344371
-376.911799817668 841.392072950387 308.701204719250
-480.299990761765 -785.093277195898 518.049893226508
-0.290993244976612 0.261509968382257 -0.555439066924084
1.22983550728053 0.353991435722233 1.13569328647925
-0.391291823344437 0.102307753059526 2.83522498748541
-2.51135997322998 -185.007116936604 776.968124044434
```

```
-0.382623084221112 -3.91936521388810 0.00343421014041810
0.00142338860603249;];

%Option 1 for the optimization of the thermodynamic
  parameters
% Para=lsqnonlin(@fun_obj,Para);
% xlswrite('clima.xls', Para);
%Option 2 for the optimization of the thermodynamic
  parameters
% Para=fminsearch(@fun_obj,Para,optimset('MaxFunEvals',500));
% xlswrite('clima.xls', Para);

[fun,Za1,Zo1,Zo1e,Za1e,Za2,Zo2,Zo2e,Za2e,Za3,Zo3,Zo3e,Za3e...
Za4,Zo4,Zo4e,Za4e,Za5,Zo5,Zo5e,Za5e]=fun_obj(Para);
figure (1)
plot(Za1e(:),Za1(:),'*g',Za2e(:),Za2(:),'*k',Zo1e(:),Zo1(:),'*g',
Zo2e(:),Zo2(:),'*k',Za3e(:),Za3(:),'*r',Zo3e(:),Zo3(:),'*r',Za
4e(:),Za4(:),'*b',Zo4e(:),Zo4(:),'*b',[0,1],[0,1],'k',[0,1.2],
[0,1],'k',[0,1],[0,1.2],'k')

figure (2)
plot(Za1e(:),Za1(:),'*b',Zo1e(:),Zo1(:),'*b',[0,1],[0,1],'k',
[0,1.1],[0,1],'k',[0,1],[0,1.1],'k')
title('AA')
figure (3)
plot(Za2e(:),Za2(:),'*r',Zo2e(:),Zo2(:),'*r',[0,1],[0,1],'k',
[0,1.1],[0,1],'k',[0,1],[0,1.1],'k')
title('Etanol')
figure (4)
plot(Za5e(:),Za5(:),'*b',Zo5e(:),Zo5(:),'*b',Za3e(:),Za3(:),'*
k',Zo3e(:),Zo3(:),'*k',[0,1],[0,1],'k',[0,1.1],[0,1],'k',[0,1]
,[0,1.1],'k')
title('Butanol')
figure (5)
plot(Za4e(:),Za4(:),'*g',Zo4e(:),Zo4(:),'*g',[0,1],[0,1],'k',
[0,1.1],[0,1],'k',[0,1],[0,1.1],'k')
title('Acetona')
```

MATLAB code to determine thermodynamic parameters

```
function[fun,Za1,Zo1,Zo1e,Za1e,Za2,Zo2,Zo2e,Za2e,Za3,Zo3,Zo3e,
  Za3e...
Za4,Zo4,Zo4e,Za4e,Za5,Zo5,Zo5e,Za5e]=fun_obj(Par)
global nnn
%
FZae=0; FZoe=0; Za=0; Zo=0;
Za(7,3)=0; Zo(7,3)=0; FZae(32,3)=0; FZoe(32,3)=0;
tol=1e-6;
Ti=298.2:5:313.2;
%Data source:
%Ghanadzadeh, H., & Ghanadzadeh, A. (2003). (Liquid+ liquid)
  equilibria in (water+ ethanol+ 2-%ethyl-1-hexanol) at
```

```
  T=(298.2, 303.2, 308.2, and 313.2) K.The Journal of Chemical
  %Thermodynamics, 35(9), 1393-1401.
FZaeo=[ not included ]; FZoeo=[ not included];
FZ=FZaeo*0.5+FZoeo*0.5;
FZae(:,1)=FZaeo(:,1); FZae(:,2)=FZaeo(:,2);
  FZae(:,3)=FZaeo(:,3);
FZoe(:,1)=FZoeo(:,1);FZoe(:,2)=FZoeo(:,2);FZoe(:,3)=FZeo(:,3);
fun(8*4*3*2,1)=0;
Vi=0.5;
T=Ti(1);
n=1;
mn=32;
mnm=0;
for i=1:32
   if i>8*n
      n=n+1;
      T=Ti(n);
   end
   zm=[1e-20 1e-20 1e-20 FZ(i,1) 1e-20 FZ(i,3) 1e-20 1e-20
      FZ(i,2)]';
   z=zm/sum(zm);
if mnm==0
Ki=[ 0.0315 0.1565 0.1794 2.0810 0.0860 0.0758 0.5501 0.147
  0.0001];
end
  [~,y,x,Ki,mnm]=ELL(z,T,tol,Ki,Vi,0,Par);
if mnm==0
    mn=mn-1;
  end
  ym=y;
  xm=x;
  Za(i,:)=[ym(4) ym(9) ym(6)];
  Zo(i,:)=[xm(4) xm(9) xm(6)];
end
Za1=Za; Zo1=Zo; Za1e=FZae; Zo1e=FZoe;
fun(1:8*4*3,1)=(FZae(:)-Za(:))*100/mn;
fun(8*4*3+1:8*4*3*2,1)=(FZoe(:)-Zo(:))*100/mn;
%%% Ethanol
Ti=298.2:5:313.2;
FZae=0; Zoe=0; Za=0; Zo=0;
Za(7,3)=0; Zo(7,3)=0;
%Data source:
%Ghanadzadeh, H., & Ghanadzadeh, A. (2003). (Liquid+ liquid)
  equilibria in (water+ ethanol+ 2-%ethyl-1-hexanol) at
  T=(298.2, 303.2, 308.2, and 313.2) K.The Journal of Chemical
  %Thermodynamics, 35(9), 1393-1401.
FZaeo=[ not included ];FZoeo=[ not included];
FZ=FZaeo*0.5+FZoeo*0.5;
FZae(28,3)=0;FZoe(28,3)=0;
FZae(:,1)=FZaeo(:,1);FZae(:,2)=FZaeo(:,2);FZae(:,3)=FZeo(:,3);
```

```
FZoe(:,1)=FZoeo(:,1);FZoe(:,2)=FZoeo(:,2);FZoe(:,3)=FZeo(:,3);
Vi=0.5;
T=Ti(1);
n=1;

mn=28;
for i=1:28
   if i>7*n
      n=n+1;
      T=Ti(n);
   end
   zm=[1e-20 FZ(i,3) 1e-20 FZ(i,1) 1e-20 1e-20 1e-20 1e-20
      FZ(i,2)]';
   z=zm/sum(zm);
   if mnm==0
 Ki=[0.0238 0.3274 0.3088 9.7510 0.0392 0.211 8.2906 0.0394
  0.0001];
   end
   [~,y,x,Ki,mnm]=ELL(z,T,tol,Ki,Vi,0,Par);
   if mnm==0
      mn=mn-1;
   end
   ym=y;
   xm=x;
   Za(i,:)=[ym(4) ym(9) ym(2)];
   Zo(i,:)=[xm(4) xm(9) xm(2)];
end
Za2=Za; Zo2=Zo; Za2e=FZae; Zo2e=FZoe;
fun(192+1:192+84,1)=(FZae(:)-Za(:))*100/mn;
fun(192+85:192+84*2,1)=(FZoe(:)-Zo(:))*100/mn;
%%% Butanol
Ti=298.2:5:313.2;
FZae=0; FZoe=0; Za=0; Zo=0;
Za(7,3)=0; Zo(7,3)=0;
%Data source:
%Ghanadzadeh, H., & Ghanadzadeh, A. (2003). (Liquid+ liquid)
  equilibria in (water+ ethanol+ 2-%ethyl-1-hexanol) at T=(298.2,
  303.2, 308.2, and 313.2) K.The Journal of Chemical
  %Thermodynamics, 35(9), 1393-1401.
FZae=[ not included]; FZoe=[ not included];
FZ=FZae*0.5+FZoe*0.5;
Vi=0.3;
T=Ti(1);
n=1;
mn=28;
for i=1:28
   if i>7*n
      n=n+1;
      T=Ti(n);
   end
```

```
   zm=[ FZ(i,3) 1e-20 1e-20 FZ(i,1) 1e-20 1e-20 1e-20 1e-20 FZ(i,2)]';
   z=zm/sum(zm);
   if mnm==0
   Ki=[0.0238 0.3274 0.3088 9.7510 0.0392 0.2117 8.2906 0.0394
    0.0001];
   end
     [~,y,x,Ki,mnm]=ELL(z,T,tol,Ki,Vi,0,Par);
     if mnm==0
     mn=mn-1;
   end
   ym=y;
   xm=x;
   Za(i,:)=[ym(4) ym(9) ym(1)];
   Zo(i,:)=[xm(4) xm(9) xm(1)];
end
Za3=Za;Zo3=Zo;Za3e=FZae;Zo3e=FZoe;
r=1;
fun(360+1:360+84,1)=(FZae(:)-Za(:))*100/mn/r^4;
r=1;
fun(360+85:360+84*2,1)=(FZoe(:)-Zo(:))*100/mn/r^4;

%%% Acetone
Ti=298.2:5:313.2;
FZae=0;FZoe=0;Za=0;Zo=0;Za(9,3)=0;Zo(9,3)=0;
%Data source:
%Ghanadzadeh, H., & Ghanadzadeh, A. (2003). (Liquid+ liquid)
  equilibria in (water+ ethanol+ 2-%ethyl-1-hexanol) at
  T=(298.2, 303.2, 308.2, and 313.2) K.The Journal of Chemical
  %Thermodynamics, 35(9), 1393-1401.
FZaeo=[ not included]; FZoeo=[ not included];
FZ=FZaeo*0.5+FZoeo*0.5;
FZae(36,3)=0;FZoe(36,3)=0;
FZae(:,1)=FZaeo(:,1);FZae(:,2)=FZaeo(:,2);FZae(:,3)=FZaeo(:,3);
FZoe(:,1)=FZoeo(:,1);FZoe(:,2)=FZoeo(:,2);FZoe(:,3)=FZoeo(:,3);
Vi=0.5;
T=Ti(1);
n=1;
mn=36;
for i=1:36
   if i>9*n
      n=n+1;
      T=Ti(n);
   end
  zm=[1e-20 1e-20 FZ(i,3) FZ(i,1) 1e-20 1e-20 1e-20 1e-20 FZ(i,2)]';
   z=zm/sum(zm);
if mnm==0
Ki=[ 0.0315 0.1565 0.1794 2.0810 0.0860 0.0758 0.5501 0.147
0.0001];
 end
```

```
  [~,y,x,Ki,mnm]=ELL(z,T,tol,Ki,Vi,0,Par);
  if mnm==0
     mn=mn-1;
  end
  ym=y;
  xm=x;
  Za(i,:)=[ym(4) ym(9) ym(3)];
  Zo(i,:)=[xm(4) xm(9) xm(3)];
end
Za4=Za;Zo4=Zo;Za4e=FZae;Zo4e=FZoe;
fun(528+1:528+108,1)=(FZae(:)-Za(:))*100/mn;
fun(528+109:528+108*2,1)=(FZoe(:)-Zo(:))*100/mn;
%%% Binary pair water-butanol
FZae=0;FZoe=0;Za=0;Zo=0;Zo(10,2)=0;Za(10,2)=0;FZae(10,2)=0;
  FZoe(10,2)=0;
Ti=[0 9.6 20.0 30.8 40.1 50.0 60.1 70.2 80.1 90.6 ];
Z=[10.33 19.0;8.98 19.7;8.03 20.1;7.07 20.6; 6.77 21.4 6.54
  22.2;6.35 24.0; 6.73 24.8; 7.04 27.4 7.26 30.6];
zm=[30 1e-20 1e-20 70 1e-20 1e-20 1e-20 1e-20 1e-20]';
zm=zm/sum(zm);
MM=[74.1 46.07 58.08 18 88.1 58.08 90.08 44 186.34]';
z=zm./MM/sum(zm./MM);
Vi=0.5;
mn=10;
for i=1:10
   T=Ti(i)+273.15;
   FZae(i,:)=[Z(i,1),100-Z(i)];
   FZae(i,:)=FZae(i,:)./MM([1 4])'/sum(FZae(i,:)./MM([1 4])');
   FZoe(i,:)=[100-Z(i,2),Z(i,2)];
   FZoe(i,:)=FZoe(i,:)./MM([1 4])'/sum(FZoe(i,:)./MM([1 4])');
   if mnm==0
 Ki=[ 0.0315 .1565 0.1794 2.0810 0.0860 0.0758 0.5501 0.147 0.0001];
   end
   [~,y,x,Ki,mnm]=ELL(z,T,tol,Ki,Vi,0,Par);
   if mnm==0
      mn=mn-1;
   end
   ym=y;
   xm=x;
   Za(i,:)=[ym(1) ym(4)];
   Zo(i,:)=[xm(1) xm(4)];
end
Za5=Za;Zo5=Zo;Za5e=FZae;Zo5e=FZoe;
fun(744+1:744+20,1)=(FZae(:)-Za(:))*100/mn;
fun(744+21:744+40,1)=(FZoe(:)-Zo(:))*100/mn;
if nnn==1
   fun=norm(((fun)));
end
display(norm(fun)/4);
```

MATLAB code to determine the objective function.

A common problem working with solvents is the binary parameters that describe the liquid–liquid equilibrium. Because of this, it is necessary to repeat an optimization process to fit all experimental data and simulation data similar to that done for the components involved in ABE fermentation considering the same objective functions and the same equations (Equations 9.2 and 9.3). For this optimization process, we used the UNIQUAC model (Abrams and Prausnitz, 1975) and the obtained binary parameters are shown in Table 9.4.

To obtain these parameters, the optimization was performed using the MATLAB code shown earlier in the text. Figure 9.6 shows the parity between

TABLE 9.4

Binary Parameters for UNIQUAC Model

Parameters	Butanol (1)-water (2)	2-Ethyl-1-hexanol (1)-water (2)	Acetone (1)-water (2)	Ethanol (1)-water(2)
A_{12}	155.31	150.949	14.865	123.9261
A_{21}	-579.36	142.459	97.472	-226.2537
B_{12}	1.0822	0	-0.019	-0.5395
B_{21}	2.5715	0	0.963	1.1212
C_{12}	-43.711e-4	0	0	4.94e-4
C_{21}	-6.77e-4	0	0	-0.0011
Ref	Winkelman et al. (2009)	Liaw et al. (2010)	Ghanadzadeh et al. (2004a, 2004b, 2004c)	Wyczesany (2010)

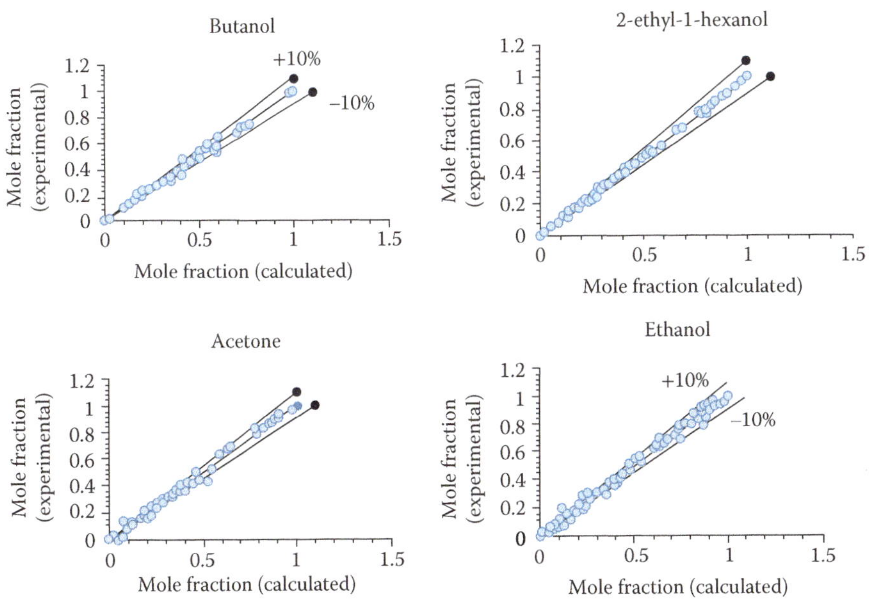

FIGURE 9.6
Parity diagram of butanol prediction by UNIQUAC model.

experimental and calculated data. Also note two lines that delimit an error region of 10%, with respect to each value. In this optimization process, almost all predicted data are inside this region. However, this behavior changes at low concentrations, but we consider that the liquid–liquid prediction is correct. These results are shown in Table 9.4.

9.4 Optimization Process

As mentioned earlier in this chapter, once we have the kinetic parameters adjusted with the experimental data, the next step is to optimize the entire process considering the integrated reactor and the separation unit (see Figure 9.3). Note we are using glucose and xylose as substrates. Energy consumption of integrated reactor depends on several reactor conditions and assumptions. For example, in liquid extraction more extractants can be used to reduce butanol concentration in the reactor and to increase the reactor productivity; however, the regeneration energy of extractant is increased.

During this optimization process, we use all these adjusted parameters and we consider liquid–liquid extraction as a separation unit and also use the previously adjusted parameters. This separation unit was used both to remove azeotropes and to recover components of interest present in the reactor. As objective function, we consider the minimization of both the total annual cost (TAC) and the eco-indicator 99 as economic and environmental indicators, respectively.

The optimization process is stated as follows:

$$\text{Min}(\text{TAC},\text{Eco99}) = \text{f}(D, E_{\text{N}}, \text{Ext}), \quad \text{s.t. } \vec{y}_k \geq \vec{x}_k \tag{9.41}$$

where D is the dilution rate (h^{-1}), E_N is the amount of enzyme added ($/kg biomass), and Ext is the amount of extractant. Several parameters used in the economic optimization are shown in Table 9.5. At a substrate cost of 0.05 $/(kg-(cellulose + xylose)), feed stream ratio of cellulose/lignin/xylose was fixed to 2/1.5/1.

The TAC has been calculated using the Guthrie method (Guthrie, 1969), as shown in Equation 9.42:

$$\text{TAC} = \frac{\text{Operating cost}(\$/\text{year}) + (\text{Total investment}(\$)/r(\text{year}))}{\text{Annual ABE production}(\text{kg}/\text{year})} \tag{9.42}$$

where r is the payback period. In this calculation, we considered 3 years.

In the optimization process, the investment cost (CI) of this process is given as follows:

$$\text{Total investment} = C_{\text{R}} + C_{\text{T}} + C_{\text{Exch}} + C_{\text{IE}} \tag{9.43}$$

where the reactor cost, column cost, condenser, and reboiler cost are represented by C_{R}, C_{T}, C_{Exch}, and C_{IE}, respectively.
The operative annual cost is calculated as follows:

$$\text{Operating cost} = C_{\text{E}} + C_{\text{V}} + C_{\text{AE}} + C_{\text{S}} + C_{\text{ENZ}} + C_{\text{Ex}} \tag{9.44}$$

where C_{E}, C_{V}, C_{AE}, C_{S}, C_{ENZ}, and C_{Ex} are the cost of electricity, steam, cooling water, substrate, enzyme, and extractant lost, respectively (see Table 9.5).

The diameter of the column and heat of the reboilers, heat exchangers, and condensers were determined using Aspen Plus®.

The environmental objective was defined in terms of eco-indicators. In particular, the impact caused by vapor on heat duty has been measured via eco-indicator 99, which is based on the methodology of the life-cycle analysis (Geodkoop and Spriensma, 2001). This eco-indicator is stated as follows:

$$\text{Eco-indicator99} = \sum_{b}\sum_{d}\sum_{k \in K} \alpha_d \omega_d \beta_b \alpha_{b,k} \tag{9.45}$$

where β_b represents the total amount of chemical b released per unit of reference flow due to direct emissions, $\alpha_{b,k}$ is the damage caused in category k per unit of chemical b released into the environment, ω_d is a weighting factor for damage in category d, and δ_d is the normalization factor for damage in category d, respectively. The values used for this calculation are reported in Table 9.6.

TABLE 9.5
Parameters Used in Economic Evaluation

Parameter	Value	Unit
Low-pressure steam	0.017	\$/MJ
Medium-pressure steam	0.022	\$/MJ
Cool water	0.06	\$/Ton
2-Ethyl-1-hexanol	4.3	\$/kg·extractant
Electricity	0.12	\$/kW·h
Enzyme	1.22	\$/kg·enzyme
Operation time (t_o)	8,500	hours
Production flow	40,000	kg·butanol
Payback period (r)	3	year

TABLE 9.6

Environmental Data for Steam

Impact categories	Vapor (points/kg)
Carcinogens	1.180E-04
Climate change	1.600E-03
Ionizing radiation	1.130E-03
Ozone depletion	2.100E-06
Respiratory effects	7.870E-07
Acidification	1.210E-02
Ecotoxicity	2.800E-03
Land occupation	8.580E-05
Fossil fuels	1.250E-02
Mineral extraction	8.820E-06

Source: B.H. Gebreslassie et al., *Appl. Energy*, 86, 1712–1722, 2009.

The multi-objective optimization approach was implemented using a hybrid platform with Microsoft Excel and Aspen Plus. The vector of decision variables (i.e., the design variables) is sent from Microsoft Excel to Aspen Plus using Dynamic Data Exchange (DDE) via COM technology. In Microsoft Excel, these values are attributed to the process variables that Aspen Plus need. After simulation, the resulting vector returns from Aspen Plus to Microsoft Excel. Finally, Microsoft Excel analyzes the values of the objective functions and proposes new values of decision variables according to the optimization method used. For the multi-objective optimization of process routes analyzed in this study, we have used the following parameters for the multi-objective differential evolution with tabu list (MODE-TL) method: 200 individuals, 500 generations, a tabu list of 50% of total individuals, and a tabu radius of 0.0000025. These nominal values are chosen based on the optimum values available in the literature (Geodkoop and Spriensma, 2001) and on the preliminary numerical experience for differential evolution (DE) and differential evolution with tabu list (DETL), 0.80 and 0.6 for crossover and mutation fractions, respectively. These parameters were obtained by a tuning process via preliminary calculations. The tuning process consists of performing several runs with different number of individuals and generations to detect the best parameters that allow us to obtain the better convergence performance of MODE-TL.

This optimization platform is shown in Figure 9.7.

9.5 Optimization Results

After the optimization process, the results obtained show a clear tendency with regard to objective function. The two objective functions are indeed in conflict with each other, which produces a behavior as shown in Figure 9.8.

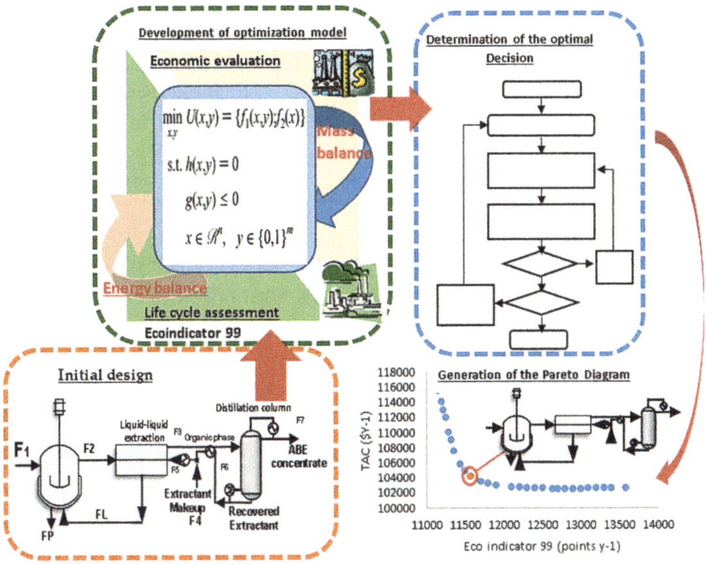

FIGURE 9.7
Optimization flowsheet for this case of study.

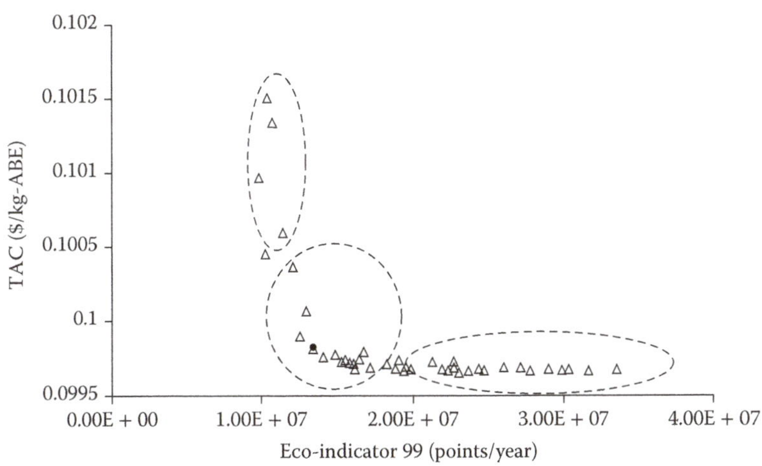

FIGURE 9.8
Pareto front facing the total annual cost (TAC) and the eco-indicator 99.

Figure 9.8 shows the Pareto fronts for two tested objective functions of all process routes. Multi-objective optimization was performed using 100,000 function evaluations. After that, the vector of decision variables did not produce a meaningful improvement. Thus, in light of this scenario, it was assumed that MODE-TL achieved the convergence at the tested numerical conditions and

TABLE 9.7

General Characteristics of Selected Designs in Pareto Front

TAC ($/kg ABE)	Eco-indicator 99 (points/year)	Dilution rate (h^{-1})	E_n amount of enzyme g-kg butanol	Ext the amount of extractant/kg butanol
0.9975	1.41×10^7	0.3055	26.2	79.81

the reported results correspond to the best solution obtained by the MODE-TL method. Each Pareto front of Figure 9.8 represents the trade-offs between both objective functions, where the x axis contains the vapor indicator (VI) values from each process and y axis represents the calculated TAC. It is clear that all process routes offered different operating conditions and significant variations on the values of optimization targets. Geometrically, three zones can be identified in each Pareto front. One section has Pareto fronts where the most expensive designs are located, but these designs represent a minor environmental impact. On the contrary, there are zones where all designs have the cheapest TAC, but the eco-indicator is the highest. At the middle of both zones, a feasible zone is located for all processes, where the obtained designs accomplish both purities and recoveries required and their TACs and eco-indicator values are in the middle of the extreme zone values. This behavior in these Pareto fronts represents the conflicting objectives along the multi-objective optimization process. For illustration, the upper zone indicated in Figure 9.8 consists of designs that include the biggest number of stages and the biggest diameter of the column but with the minor heat duty. These combinations produced the biggest TAC but the smallest eco-indicator. On the other hand, the lower zone in the Pareto front consists of designs that include the minor number of stages, the smaller diameters of column but the highest heat duty, which produced the minor TAC but the highest value of eco-indicator. Finally, the middle zone in the Pareto front includes the designs with average values of both TAC and VI; obviously, this green zone represents all feasible designs where both targets are accomplished in the best manner. In this zone, we selected a design that shows TAC of 0.9975 $/kg ABE and an eco-indicator of 1.41×10^7 points/year.

The stream leaving the reactor has a concentration of 6.77 g/L butanol with a total feed flow to the recovery column of 1270.91 kmol/h at the optimum point. The column has 24 plates, reflux ratio of 5.18, feed stream location at stage 13, and a diameter of 0.81 m (see Table 9.7).

9.6 Conclusions

An integrated reactor made possible to improve the production of biobutanol by fermentation. The optimization applied to process design is a very useful tool to obtain design, which accomplishes several objective functions such as economic and environmental. This chapter shows that this tool helps improve the thermodynamic parameters with the experimental data. Consequently, all simulation results will be closer to the real process. Furthermore, the optimization process improves the economic and environmental indicators or even other objective functions according to the needs of designer.

References

N. Abdehagh, F.H. Tezel, J. Thibault, 2014, Separation techniques in butanol production: Challenges and developments. *Biomass Bioenergy*, 60, 222–246.

D.S. Abrams, J.M. Prausnitz, 1975, Statistical thermodynamics of liquid mixtures: A new expression for the excess Gibbs energy of partly or completely miscible systems. *AIChE J.*, 21, 116–128.

P. Andrić, A.S. Meyer, P.A. Jensen, K. Dam-Johansen, 2010, Reactor design for minimizing product inhibition during enzymatic lignocellulose hydrolysis: II. Quantification of inhibition and suitability of membrane reactors. *Biotechnol. Adv.*, 28, 407–425.

M. Chen, F. Wang, 2010, Optimization of a fed-batch simultaneous saccharification and cofermentation process from lignocellulose to ethanol. *Society*, 1400, 5775–5785.

V. H. G. Díaz, G. O. Tost, 2016, Butanol production from lignocellulose by simultaneous fermentation, saccharification, and pervaporation or vacuum evaporation. *Bioresour. Technol.*, 218, 174–182.

B.H. Gebreslassie, G. Guillen-Gosalbez, L. Jimenez, D. Boer, 2009, Design of environmentally conscious absorption cooling systems via multiobjective optimization and life cycle assessment. *Appl. Energy*, 86, 1712–1722.

M. Geodkoop, R. Spriensma, 2001, The eco-indicator 99. A damage oriented for life cycle impact assessment. Methodology report and manual for designers. Technical Report, Amersfoort, the Netherlands: PRe´ Consultants.

H. Ghanadzadeh, A. Ghanadzadeh, 2003, (Liquid+liquid) equilibria in (water+ethanol+ 2-ethyl-1-hexanol) at T=(298.2, 303.2, 308.2, and 313.2) K. *J. Chem. Thermodyn.*, 35, 1393–1401.

H. Ghanadzadeh, A. Ghanadzadeh, 2004a, Liquid–liquid equilibria of water + 1-butanol + 2-ethyl-1-hexanol system. *Engineering*, 783–786.

H. Ghanadzadeh, A. Ghanadzadeh, M. Alitavoli, 2004b, LLE of ternary mixtures of water/acetone/2-ethyl-1-hexanol at different temperatures. *Fluid Phase Equilibr.*, 219, 165–169.

H. Ghanadzadeh, A. Ghanadzadeh, R. Sariri, 2004c, (Liquid+liquid) equilibria for (water+acetic acid+2-ethyl-1-hexanol): Experimental data and prediction, *J. Chem. Thermodyn.*, 36, 1001–1006.

H. González-Peñas, T.A. Lu-Chau, M.T. Moreira, J.M. Lema, 2014, Solvent screening methodology for in situ ABE extractive fermentation. *Appl. Microbiol. Biotechnol.*, 98, 5915–5924.

K.M. Guthrie, 1969, Capital cost estimating, *Chem. Eng.*, 76, 114.

Y. Jang, A. Malaviya, C. Cho, J. Lee, S. Lee, 2012, Butanol production from renewable biomass by clostridia. *Bioresour. Technol.*, 123, 653–663.

D.T. Jones, D.R. Woods, 1986, Acetone–butanol fermentation revisited. *Society*, 50, 484–524.

S. Junne, 2010, *Stimulus Response Experiments for Modelling Product Formation in Clostridium acetobutylicum Fermentations,* Berlin: Technical University of Berlin.

K.L. Kadam, E.C. Rydholm, J.D. McMillan, 2004, Development and validation of a kinetic model for enzymatic saccharification of lignocellulosic biomass. *Biotechnol. Progr.*, 20(3), 698–705.

R.D. Li, Y.Y. Li, L.Y. Lu, C. Ren, Y.X. Li, L. Liu, 2011, An improved kinetic model for the acetone–butanol–ethanol pathway of Clostridium acetobutylicum and model-based perturbation analysis. *BMC Syst. Biol.*, 5, S12.

H.-J. Liaw, V. Gerbaud, C.-C. Chen, C.-M. Shu, 2010, Effect of stirring on the safety of flammable liquid mixtures. *J. Hazard. Mater.*, 177, 1093–1101.

N. Qureshi, H.P. Blaschek, 2001, Recovery of butanol from fermentation broth by gas stripping. *Renew. Energy*, 22(4), 557–564.

N. Qureshi, V. Singh, S. Liu, T.C. Ezeji, B.C. Saha, M.A. Cotta, 2014, Process integration for simultaneous saccharification, fermentation, and recovery (SSFR): Production of butanol from corn stover using *Clostridium beijerinckii* P260. *Bioresour. Technol.*, 154, 222–228.

R. Rathmann, A. Szklo, R. Schaeffer, 2010, Land use competition for production of food and liquid biofuels: An analysis of the arguments in the current debate. *Renew. Energy*, 35, 14–22.

H. Shinto, Y. Tashiro, M. Yamashita, G. Kobayashi, T. Sekiguchi, T. Hanai, et al., 2007, Kinetic modeling and sensitivity analysis of acetone–butanol–ethanol production. *J. Biotechnol.*, 131, 45–56.

H. Shinto, Y. Tashiro, G. Kobayashi, T. Sekiguchi, T. Hanai, Kuriya Y, et al., 2008, Kinetic study of substrate dependency for higher butanol production in acetone–butanol–ethanol fermentation. *Process Biochem.*, 43, 1452–1461.

G. Sorda, M. Banse, C. Kemfert, 2010, An overview of biofuel policies across the world. *Energy Policy*, 38, 6977–6988.

J.G.M. Winkelman, G.N. Kraai, H.J. Heeres, 2009, Binary, ternary and quaternary liquid–liquid equilibria in 1-butanol, oleic acid, water and n-heptane mixtures. *Fluid Phase Equilib.*, 284, 71–79.

A. Wyczesany, 2010, Calculation of vapour–liquid–liquid equilibria. *Chem. Process Eng.*, 31, 333–353.

V.V. Zverlov, O. Berezina, G.A. Velikodvorskaya, W.H. Schwarz, 2006, Bacterial acetone and butanol production by industrial fermentation in the Soviet Union: Use of hydrolyzed agricultural waste for biorefinery. *Appl. Microbiol. Biotechnol.*, 71, 587–597.

10

Optimization of a Silane Production Process*

10.1 Introduction

The photovoltaic solar energy, like many other renewable energy resources, constitute an inexhaustible source of energy facing the fossil fuels, contributing the self-supplying national energy and the social sector, with a comparative environmental impact considerably smaller than that of the conventional source of energy.

At a global level, with the reduction in the stock of fossil fuels (particularly petroleum) and the urgent necessity to count with alternative energy sources, preferently renewable, clean and economically viable energy sources, with eolic energy, hydroelectric power and solar energy taking big relevance these days. The search of these alternative sources is one of the most important challenges that humanity faces nowadays. In particular, solar energy is being explored in many different ways, but the most popular method includes solar cells based on silicon, which transform sunlight directly into electricity through the photovoltaic effect.

The growing interest in the production of silicon as a raw material for manufacturing solar cells has evolved in an important way in the last decades (see Table 10.1), and an increase in demand up to around 30% is expected (Braga et al., 2008). A more detailed analysis remarks that the photovoltaics market has grown up at an average speed of 45% per year during the last decade, but with a remarkable increase of approximately 70% per year between 2007 and 2011, and a considerable decrease of 15% in 2012 because of reduction in the economic stimulus of some European countries. Meanwhile, the numbers speak of a huge increase; the installed capacity of 27 GW in 2011 is just a fraction that represents 1% of the total energy quantity generated by all means and indicates that there is much scope to continue development on the immediate future (Zweibel et al., 2008; Wolden et al., 2011).

Even when silicon cells are competing with other type of cells manufactured using advanced materials, it has been predicted that silicon solar cells will continue to significantly contribute to the market as the technology matures, of their availability, and mainly if advancement in the present technology leads to reduction in their cost (Morales-Acevedo and Casados-Cruz, 2013).

* César Ramírez-Márquez, Eduardo Sánchez-Ramírez and Juan José Quiroz-Ramírez also contributed to the work on this chapter.

TABLE 10.1

Demand and Availability of Polycrystalline Silicon in the Last Decade

Year	Capacity (*t*)	Demand		Availability
		Semiconductor (*t*)	Photovoltaic (*t*)	Photovoltaic (*t*)
2003	26,700	17,000	9,000	9,700
2004	28,800	19,350	14,032	9,450
2005	30,200	20,085	18,181	10,115
2006	34,500	21,166	16,705	13,334
2007	38,050	23,071	17,435	14,979
2008	48,550	26,301	24,089	22,249
2009	53,800	26,837	28,233	26,973
2010	5,800	27,632	32,108	31,168

Source: A.F.B. Braga et al., *Sol. Energy Mater. Sol. Cells.*, 92, 418, 2008.

Polycrystalline cells, monocrystalline cells, and amorphous silicon are the first most commonly used solar cells. However, it is important to know that the high cost of polycrystalline silicon is due to higher production cost for the process itself and due to the production of its raw material, silane (SiH_4) (Pizzini et al., 2005).

One of the processes developed many years ago but which is still valid for the production of silane is disproportionation of trichlorosilane into silane and silicon tetrachloride (i.e., the reaction between metallurgical grade silicon and hydrogen chloride; metallurgical grade silicon is a result of the carbothermic reduction of mineral quartz) (Bakay, 1976). It has been analyzed that approximately 40% of the energy required for fabrication of a silicon-based solar panel is consumed in the production of its precursor. That is why, the reduction in energy consumption at this stage is crucial to minimize the return module and with that the technology cost (Werner et al., 2013).

The reactions of the Si–H–Cl system are considerably complex, and there are only a few studies that have reported the so-called Siemens process (a process in which the silicon is formed through silane) with calculations based on the thermodynamics data selection (Jun-Jian et al., 2007). The production of silane is, with no doubt, a process that involves numerous issues, mainly on the production cost because of the high purity that is required. An alternative to the conventional process of silane production is reactive distillation (RD), which overcomes the disadvantage of the conventional process: this last one requires multiple distillation columns to obtain the final product.

Having said that, we present a proposal of conceptual design for a unique RD column to produce high-purity silane, dichlorosilane, and monochlorosilane—products of great commercial and industrial interest—based on the principles of the process intensification strategies, aiming to recuperate the products of interest and to minimize environmental impact.

10.2 Silane Production

The conventional process for silane production by the proportional decomposition of trichlorosilane consists of two reactors and multiple distillation columns (Coleman, 1982; Breneman, 1983). Figure 10.1 shows the typical process, which can be used as a base case scenario. In this case, the first reactor is used for the first proportional decomposition reaction of trichlorosilane to dichlorosilane and the second reactor is used for combining both to give way to succeeding proportional decomposition reactions. Besides, four distillation columns are used to separate the products starting with the reactives that later will be recycled.

One of the disadvantages of such conventional processes is that the recycling rates are extremely high. For example, for a production process of silicon with 5000 ton/year capacity, the recycling of trichlorosilane need to be done at the rate of approximately 94 ton/year. A process with such recycling rates calls for huge equipment and very high energy costs (Muller et al., 2002).

In the conventional process of silane, dichlorosilane, and monochlorosilane production, the necessity of high recycle rates arises from the unfavorable chemical equilibrium. This conventional process requires high

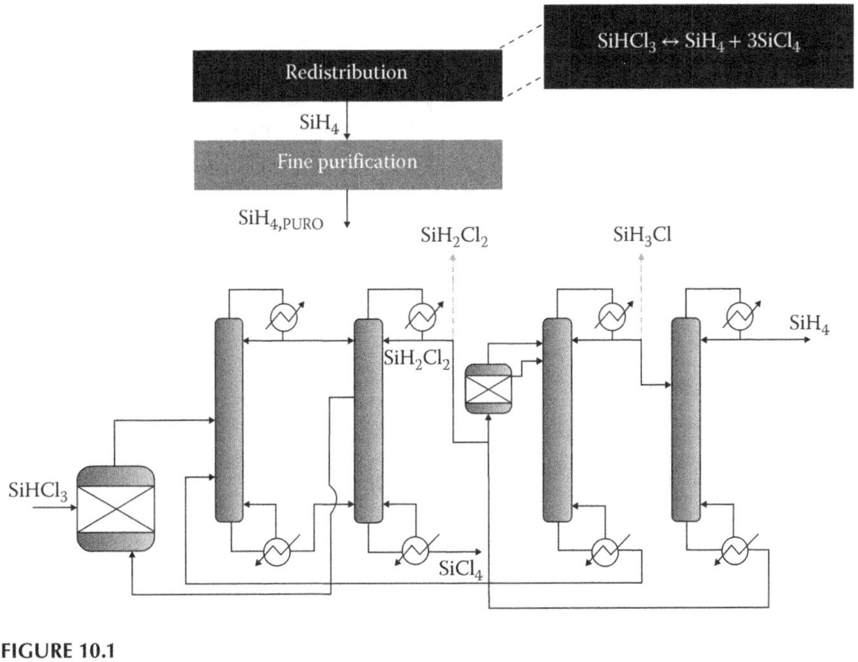

FIGURE 10.1
Conventional production of silane. (From M. Yamada et al., *Method for Continuous Production of Silanes*, Denki Kagaku Kogyo Kabushiki Kaisha, JP121110, 1984.)

capital and has high operational cost, which is due to the high recycling rates as mentioned earlier.

10.3 Description of the Process Using Reactive Distillation

Reactive distillation is a technology particularly attractive for reactions such as proportional decomposition of trichlorosilane. The basic idea, as mentioned in Chapter 8, is to combine the reaction and distillation into a single column.

Irrespective of the described advantages of RD column for proportional decomposition of trichlorosilane, there is a series of possible limitations and difficulties (Muller et al., 2002). General limitations and their consequences on silane production are detailed in the following paragraphs.

First, the relative volatility of the components should be checked. The reactants and products must have the required volatility to maintain high concentrations of products in the reaction zone. Figure 10.2 shows the boiling points of silane, trichlorosilane, and silicon tetrachloride at a pressure of 5 bar. The differences in boiling point are very high, therefore separation by distillation in these cases is very simple. Hence, the relative volatility is favorable for the application of RD.

Furthermore, high reaction rates can be obtained under distillation conditions. Here, it is important to note that the proportional decomposition reactions should take place in a small temperature interval. If the temperature is too low, then the reactions are slow and their residence time is high. On the other hand, the high temperature results in undesired catalyst deactivation rates. The optimum temperature range depends on the catalyst. Experimental data show that reasonable reaction rates are reached at temperatures below 100°C, which can be easily achieved in the distillation columns operating between 1 and 10 bar (Muller et al., 2002).

Brutto reaction:	$4\ SiHCl_3$	=	SiH_4	+	$3\ SiCl_4$	
Boiling point at 5 bar:	87.2°C	=	−77.7°C	+	116.8°C	

Chemical equilibria at 80°C

$2\ SiHCl_3$	↔	SiH_2Cl_2	+	$SiCl_4$	A 9.8 mol% SiH_2Cl_2
$2\ SiH_2Cl_2$	↔	SiH_3Cl	+	$SiHCl_3$	A 28 mol% SiH_3Cl
$2\ SiH_3Cl$	↔	SiH_4	+	SiH_2Cl_2	A 36 mol% SiH_4

FIGURE 10.2
Chemical balance of proportional decomposition of trichlorosilane.

Next are the residence time limitations. If the residence time for the reaction is long, a large column size and high retention values are required. Thus, a conventional reactor-separator adjustment could be more economic. However, a detailed design requires kinetic data and thus a kinetic study for the selected catalyst under distillation conditions.

A final restriction is a result of catalyst deactivation. Based on the column configuration, use of packed catalyst for RD may not be suitable. It is important to have the possibility of substituting catalyst in an easy way so that it remains economic.

Considering all the aforementioned points, RD is found to be technically possible and attractive technology for the production of silane, dichlorosilane, and monochlorosilane. The first RD process for the proportional decomposition reactions of trichlorosilane in a fixed-bed reactor consisting of a solid ion-exchange resin was proposed by Bakay (1976). Yamada et al. (1984) proposed a process using dimethylamine or diethylamine (medium-boiling-point components) as homogeneous catalyzer in the interior of the column. Inoue (1988) developed an RD process for proportional decomposition of dichlorosilane to silane, trichlorosilane, and silicon tetrachloride. Matthes et al. (1988) and Frings (2000) published a distillation process with a reactor next to the proportional decomposition of trichlorosilane to dichlorosilane and silicon tetrachloride.

Huang et al. (2013) published a reactive distillation process that combines the three reactions of proportional decomposition into a single column with intermediate condensers for silane purification. Figure 10.3 shows a simplified flow diagram of the reaction column that includes three sections: one reaction section in the middle of the column and two separation sections, one below (stripping section) and another above (rectifying section) the

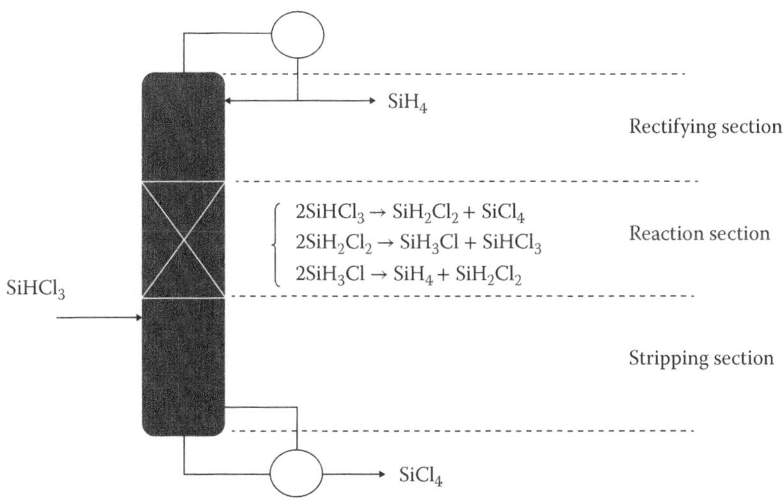

FIGURE 10.3
Reactive distillation process for silane production.

reaction zone. Trichlorosilane is supplied to the section below the reaction column. The products of this reaction are silicon tetrachloride at the bottom and silane in the dome. Silicon tetrachloride is a compound with a high boiling point, which is purified in the exhaust section and is withdrawn as a product in the tails. Dichlorosilane breaks down inside the reaction section: first as monochlorosilane and then as silane. Finally, silane, which is the desired product, is purified in the rectification section and is withdrawn as a distillate. An important point in this study is the use of an intermediate condenser, which removes most of the condensation heat. Also, the use of an intermediate condenser has a significant influence on the process economy.

Huang et al. (2013) showed the feasibility of obtaining silane with a purity of more than 99% and a full conversion of trichlorosilane to silane and silicon tetrachloride using a common RD column. Owing to extremely low condensation temperature (approximately 80°C) and excessively high reflux ratio (more than 60) in the superior part of the column, an intermediate condenser was introduced between the rectification and the reaction sections to reduce the refrigeration requirement. Using this procedure, the reflux ratio was reduced by more than 1 unit, which produces a 97% reduction in the refrigeration load. A second intermediate condenser was attached inside the reaction section to provide an additional 50% reduction in the condensation load. As a result, the global economy of the process can be significantly improved.

10.4 Economic Potential of Reactive Distillation Production of Silane

Figure 10.4 shows the estimated production costs for the RD process based on a column design for catalytic packing. The cost data are presented in comparison with the corresponding cost data from the conventional silane process that is shown in the figure. The capacity base is 5000 t/year for pure silicon (Muller et al., 2002).

This comparison shows a significant reduction in cost, and can be achieved using the RD column. Two reasons for this reduction in cost are as follows:

1. Equipment cost is reduced by more than 45%. As a consequence, depreciation and maintenance costs are reduced by approximately 45%.
2. Energy cost is reduced clearly by 60%.

10.5 Thermodynamics and Kinetic Model

The reaction consists of three steps. First, trichlorosilane ($SiHCl_3$) is converted into dichlorosilane (SiH_2Cl_2) and silicon tetrachloride ($SiCl_4$). Second,

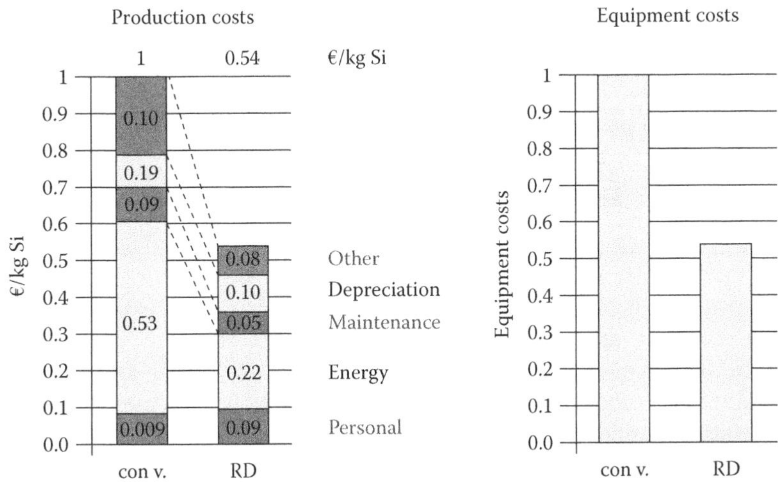

FIGURE 10.4
Economic comparison of reactive distillation process and conventional process for silane production: Estimated production and equipment costs. (From M. Yamada et al., *Method for Continuous Production of Silanes*, Denki Kagaku Kogyo Kabushiki Kaisha, JP121110, 1984.)

dichlorosilane reacts to obtain monochlorosilane (SiH_3Cl) and trichlorosilane. Finally, monochlorosilane is transformed into silane (SiH_4) and dichlorosilane. The reaction steps are shown by Equations 10.1 through 10.3:

$$2SiHCl_3 \rightarrow SiH_2Cl_2 + SiCl_4 \tag{10.1}$$

$$2SiH_2Cl_2 \rightarrow SiH_3Cl + SiHCl_3 \tag{10.2}$$

$$2SiH_3Cl \rightarrow SiH_4 + SiH_2Cl_2 \tag{10.3}$$

The reaction velocities are shown by Equations 10.4 through 10.6:

$$r_1 = k_1\left(x_1^2 - \frac{x_0 x_2}{K_1}\right) \tag{10.4}$$

$$r_2 = k_2\left(x_2^2 - \frac{x_1 x_3}{K_2}\right) \tag{10.5}$$

$$r_3 = k_3\left(x_3^2 - \frac{x_2 x_4}{K_3}\right) \tag{10.6}$$

where x_0, x_1, x_2, x_3, and x_4 are the molar fraction of silicon tetrachloride, trichlorosilane, dichlorosilane, monochlorosilane, and liquid silane, respectively; r_1, r_2, and r_3 are the reaction rate of trichlorosilane, dichlorosilane, and unproportional monochlorosilane, respectively; and k and K are the

TABLE 10.2

Kinetic Parameters for the Disproportionation of Trichlorosilane in the Liquid Phase

	k_0 (s^{-1})	E (J/mol)	K_0	ΔH (J/mol)
r_1	73.5	30,045	0.1856	6,402
r_2	949466.4	51,083	0.7669	2,226
r_3	1176.9	26,320	0.6890	−2,548

Source: X. Huang et al., *Ind. Eng. Chem. Res.*, 52, 6211, 2013.

reaction constant and the chemical equilibrium constant, respectively (see Table 10.2).

A suitable thermodynamic model is as relevant as the kinetic model in the simulation of an RD column, given this process is based on a simultaneous reaction and separation of phases. In this study, provided the mix is nonpolar, the incorporated model Peng–Robinson equation was selected to achieve the thermodynamic calculus. The proportional decomposition of trichlorosilane can be carried out by several types of catalysts (Huang et al., 2013), such as aluminum trichloride, and some ion-exchange resins. Among these catalysts, amine ion exchange is the best option because of its relatively high activity and immobilized solid state. In this study, the catalyst "Amberlyst" (A-21) was selected because it has reasonable reaction rates between 30°C and 80°C, with a suggested operative thermal resistance of 100°C.

10.6 Initial Design

Huang et al. (2013) provided an RD process that combines the three proportional decomposition reactions (Equations 10.1 through 10.3) into a single column with intermediate condensers for silane purification. The products were silicon tetrachloride and silane, which were obtained at the bottom and in the dome of the column. Silicon tetrachloride is a compound with a high boiling point; it was purified in the stripping section and the purified product was obtained at the tail. Dichlorosilane proportionally decomposes in the reaction section: first as monochlorosilane and then as silane. Finally, silane, which is the desired product, was purified in the rectification section and was withdrawn as distillate.

The simulation was performed using the rigorous model "RadFrac" with the commercial simulator Aspen Plus® to implement the steady-state calculations that imply the proportional decomposition of trichlorosilane into the components, the kinetic and the phase equilibrium models. To simplify, the reactions were specified as pseudo-homogeneous in the liquid phase. The goal of the simulation was to obtain silane with high purity, more than 99% (mol), to meet, at least, the solar-grade silane requirement.

The RD column comprised 60 stages. The reaction section was from stage 16 to stage 45, with a residence time of 2.5 s for each stage, known as retention. The column makes it function with a high pressure of 5 atm and with a pressure decrease of 0.5 kPa at each stage. The distillate–feed ratio is fixed at 0.25 mol. Trichlorosilane is supplied to the column at 10 kmol/h; this occurs in stage 46 under the temperature and pressure conditions of 50°C and 5.5 atm, respectively.

10.6.1 Buildup of the Initial Column

To start the simulation, the software Aspen Plus V. 8.4 is first initialized. In the window "Start using Aspen Plus", the "New" option is selected. Then, "Blank Simulation" is selected and then the "Create" button is pushed. In the lower part, we find three options: "Properties," "Simulation," and "Energy Analysis"; by default, the program shows the "Properties" option and the tab "Selection." In this tab, we introduce all the components that are required for all the process to take place. The components, as shown in Figure 10.5, are trichlorosilane ($SiHCl_3$), tetrachlorosilane ($SiCl_4$), dichlorosilane (SiH_2Cl_2), silane (SiH_4), and monochlorosilane (SiH_3Cl).

Then, the thermodynamic model suitable for all the components was selected. In this study, the "Peng–Robinson" method was selected, as the literature indicates that this method better approximates the experimental parameters (Huang et al., 2013). Figure 10.6 indicates where to introduce the mentioned thermodynamic model. When all the required information is completed, such as the components and the thermodynamic model, the simulation is run and the "Properties" section is completed.

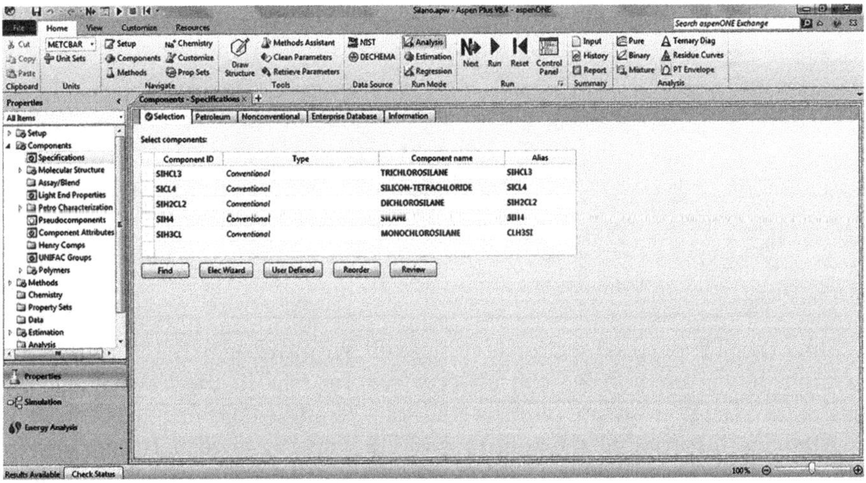

FIGURE 10.5
Window to introduce all the components.

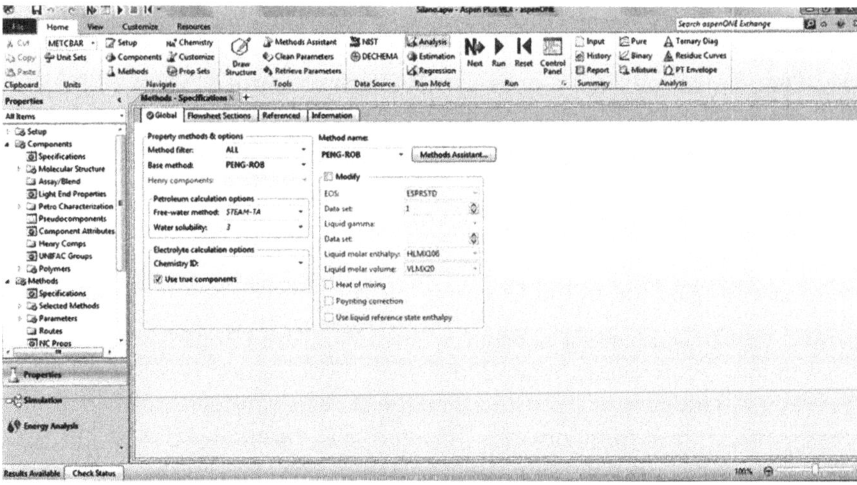

FIGURE 10.6
Window to introduce the thermodynamic model.

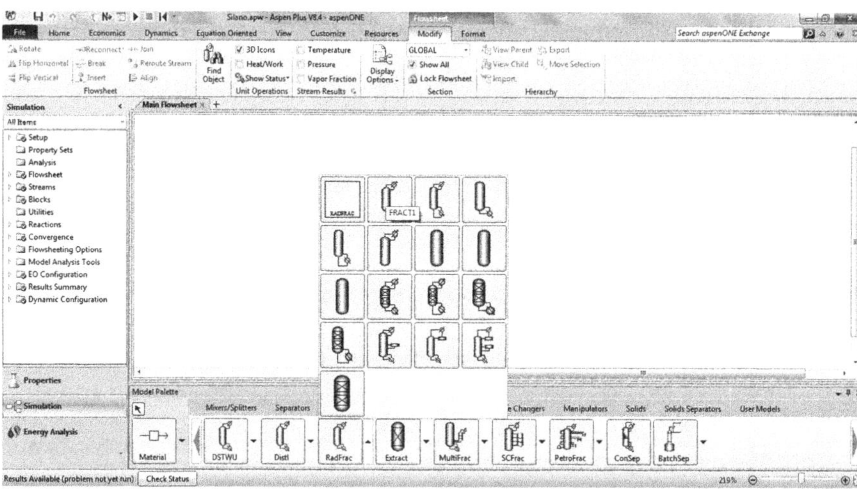

FIGURE 10.7
Window to introduce the column "RadFrac."

Once the "Properties" section is completed, we move on to the "Simulation" section. In Figure 10.7, we can observe several tabs in the lower part. We select the part that says "Columns-RadFrac" to add a column.

After we incorporate the column, we click on the lower left part to add the current lines located in the "Material" module, as shown in Figure 10.8. Therefore, the following will be added: the feed stream (F), the distillate stream (D), and the bottoms stream (B).

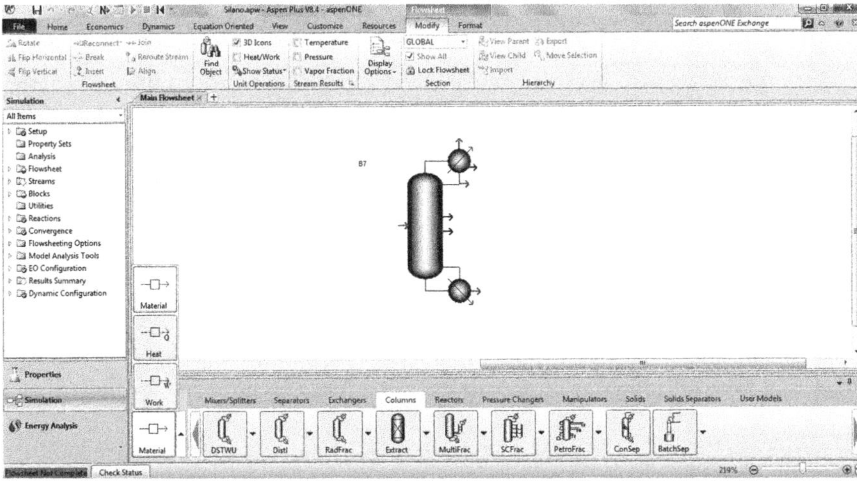

FIGURE 10.8
Window to introduce the column flow lines.

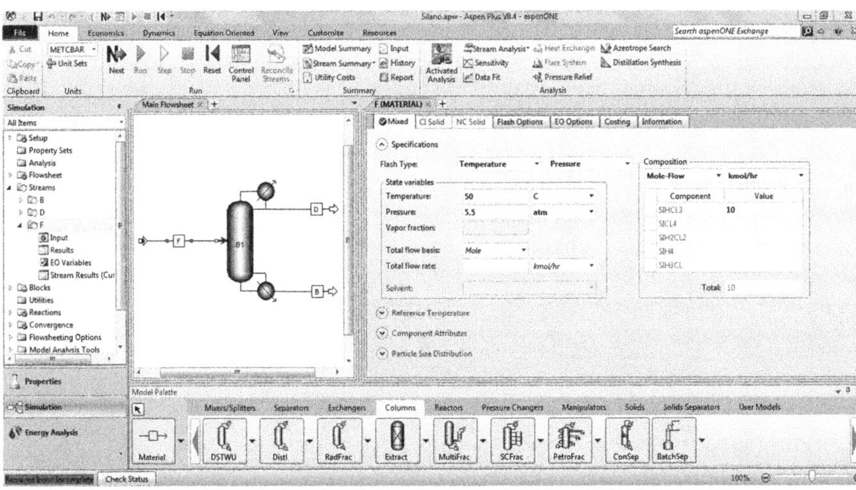

FIGURE 10.9
Window to introduce the feed current data (F).

Once the column streams are set up in the column, we introduce data for the feed stream (F): temperature (50°C), pressure (5.5 atm), and feed flow (10 kmol/h of trichlorosilane); see Figure 10.9.

Next, we double click the column to introduce the parameters that require the "Configuration" option where the number of stages (in this study case, it is 60) will be required. It is important to leave the default options provided by the program in the "Condenser, Reboiler, Valid phases, and Convergence" section. In the lower part, the operation specification is set. "Reflux ratio" and "Distillate to feed ratio" are selected as variables (63 and 0.25, respectively), as shown in Figure 10.10.

Now, the "Streams" and "Pressure" tabs should be filled; in the "Streams" option, the feed stage, which is set as 46 (the feed stage with the configuration "On Stage") should be filled. In the "Pressure" option, we fix the operation pressure as 5 atm and the pressure drop as 0.5 kPa, as shown in Figures 10.11 and 10.12.

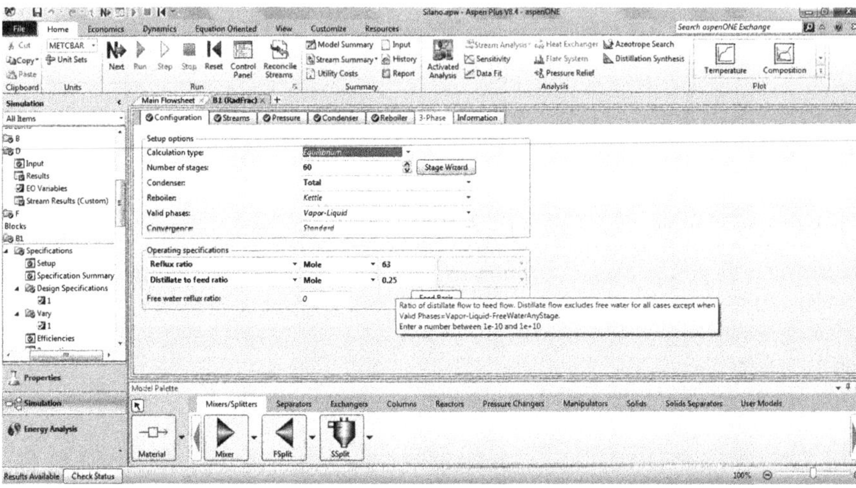

FIGURE 10.10
Window to introduce the window data.

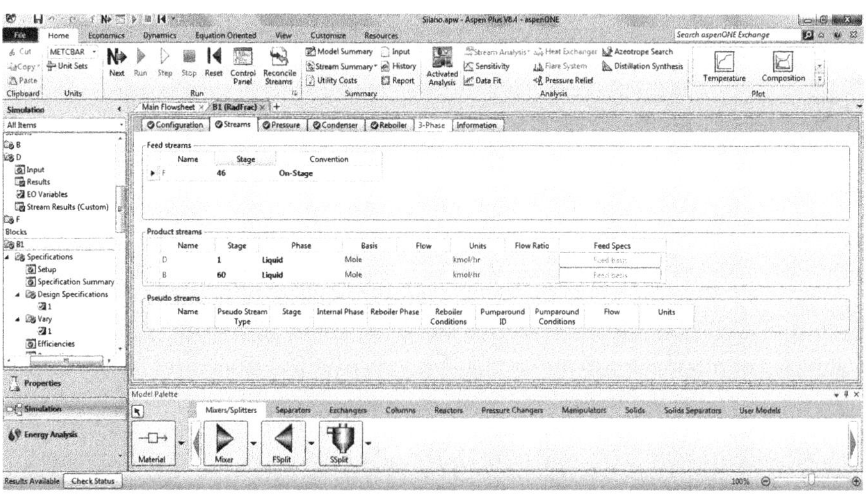

FIGURE 10.11
Windows to introduce the streams of the column data.

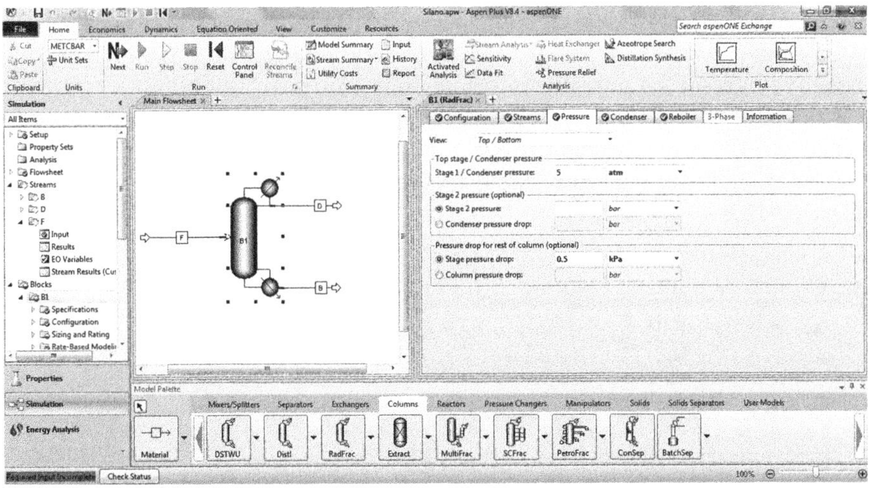

FIGURE 10.12
Window to introduce the pressure of the column data.

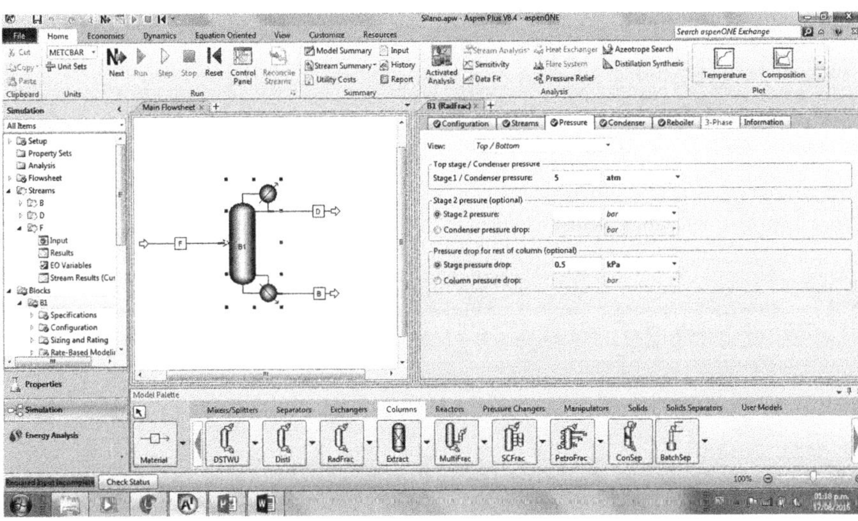

FIGURE 10.13
Window used to introduce the reaction information.

An extremely important part is to insert the reaction and the necessary parameters for the kinetic part. On the left side of the window, we can see several folders. We double click the folder named "Reactions" and insert a new reaction (R1), as shown in Figure 10.13.

The previously defined reactions are introduced into the "Stoichiometry" tab with the option "New." In the window "Edit Reaction," we select the upper right part "Reaction Type" and modify the "Equilibrium" data to "Kinetic" for each reaction that we insert, along with the reactants and the products with their respective coefficients. The reactant coefficients should be inserted with a negative sign; all the previous information is shown in Figure 10.14.

Consequently, it is essential to introduce the kinetic parameters of each reaction—the pre-exponential (*k*) and the activation energy (*E*)—in the "Kinetic" tab; the base on "Mole Fraction," should be fixed, as shown in Figure 10.15.

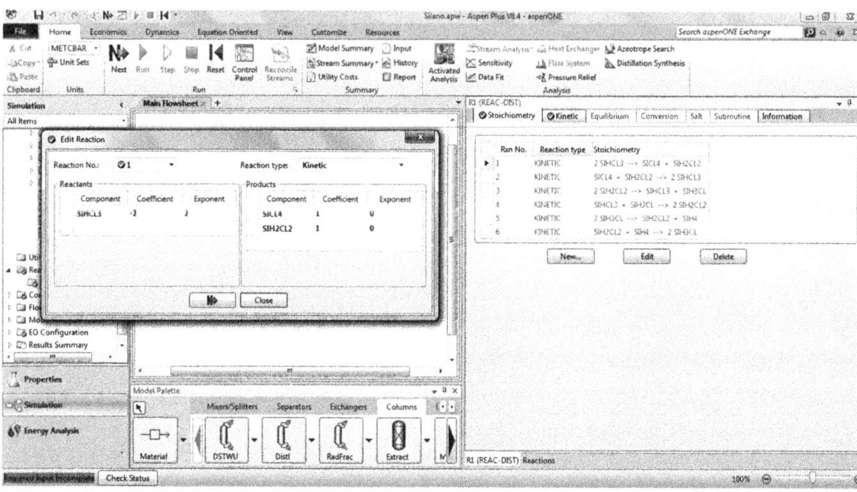

FIGURE 10.14
Window to introduce the reaction coefficients.

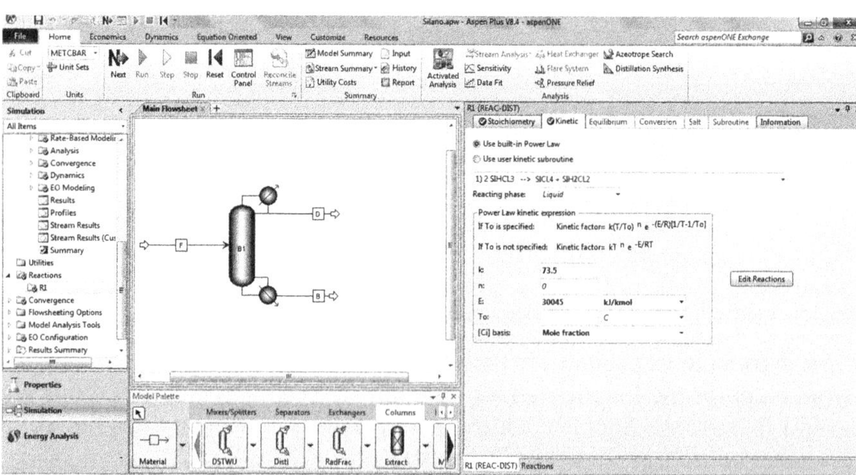

FIGURE 10.15
Window to insert the reaction kinetic parameters.

In Figure 10.16, we present and introduce all the reactions that are essential for the reactive distillation process.

Now, it is necessary to define the section of stages of the column that is to be denominated as Reactive stages. In the folder located on the left side of the window, it is "Specifications" from block B1; we choose the "Reactions" option to insert the initial stage or "Starting Stage" and the final stage or "Ending Stage," which correspond to 16 and 45, respectively, which is similar to that we selected for the reaction in "Reaction ID," as shown in Figure 10.17.

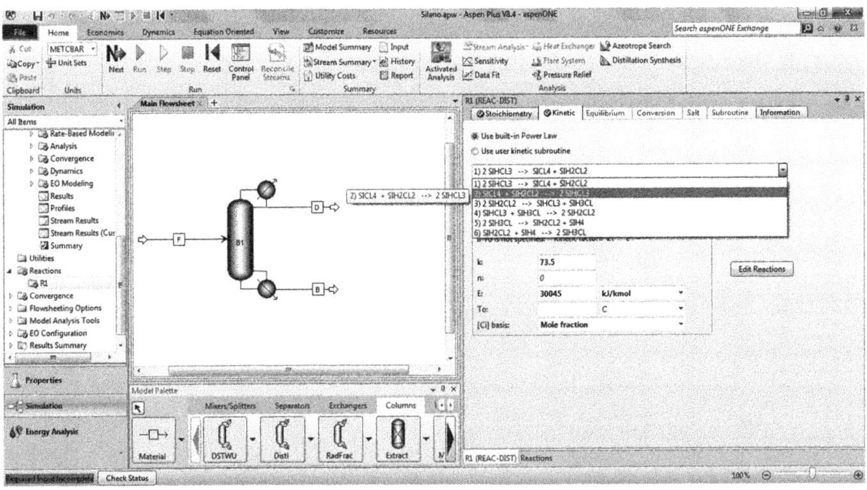

FIGURE 10.16
Window used to put on all the reactions.

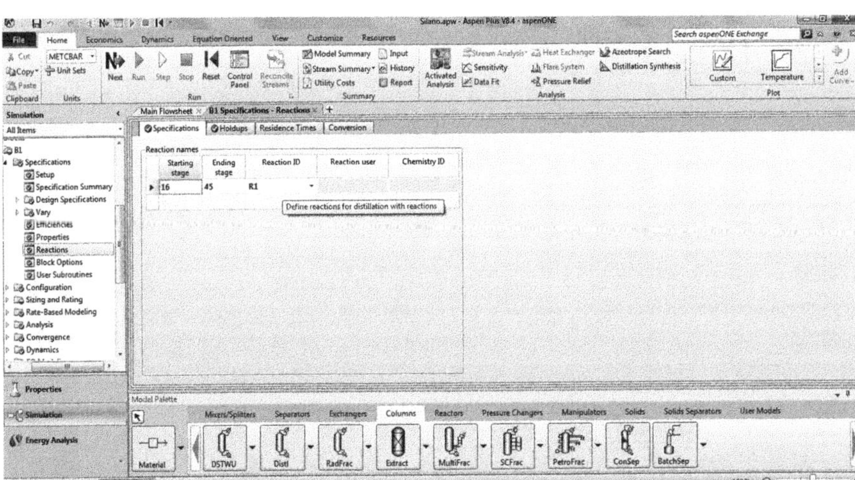

FIGURE 10.17
Window to set up the reactive stages.

Moreover, it is essential to establish the residence time for the reaction to take place; it is fixed in the liquid phase as 2.5 s for each stage, as shown in Figure 10.18.

Next, we introduce the design specifications to achieve the purity requirement of 0.995 mol fraction. In the same folder of "Specifications" of the block B1, we select the "Design Specifications" option and add a new one, and on the presented tab called "Specifications," in the upper part, we select the "Mole Purity" option and set the target as 0.995 (see Figure 10.19). To conclude, this specification is

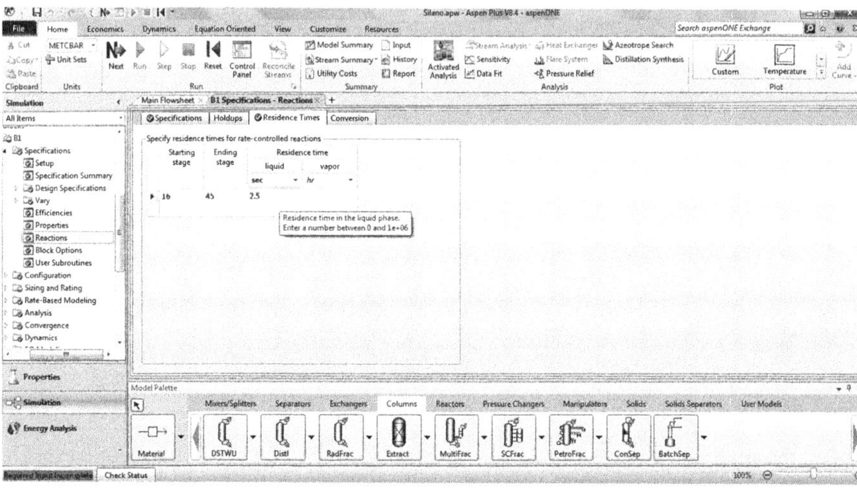

FIGURE 10.18
Window to set up the residence time.

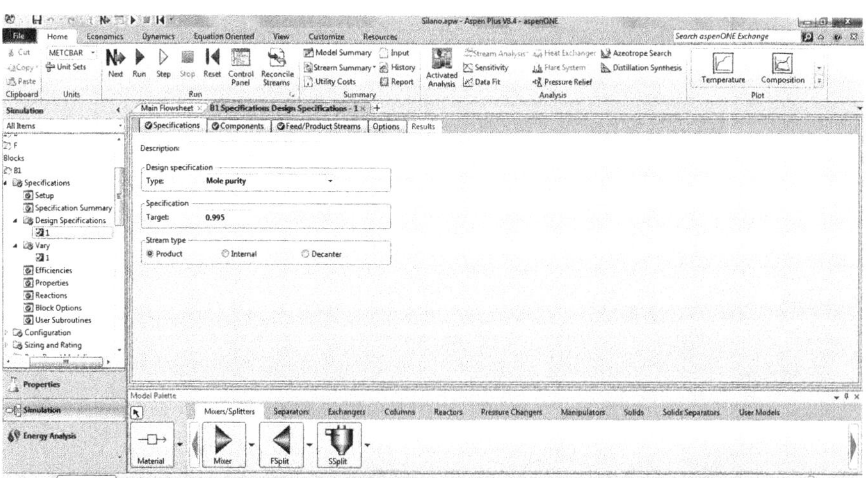

FIGURE 10.19
Window of selection for the purity goal on "Design Specification."

necessary to select the component of interest on the "Components" tab and passing it to the right side of the white baskets with the corresponding arrow direction. In this case, the selected component is silane, as shown in Figure 10.20. Finally, we choose the stream to achieve our product of interest by selecting the "Feed/Product Stream" tab, in this case, by the distillate, stage 1, as shown in Figure 10.21.

To complete the analysis, it is essential to indicate the variable that will be manipulated to find the required purity. Therefore, we select the folder "Vary" and the variable "Reflux Ratio" because it is the variable to be

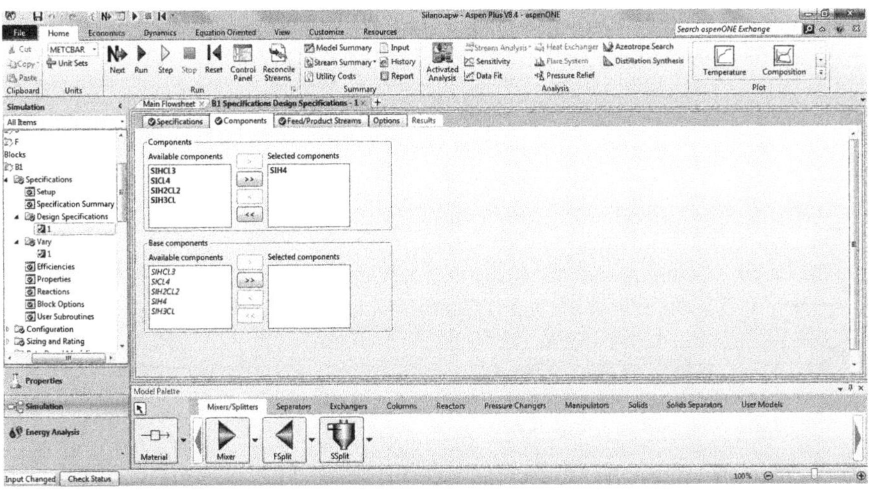

FIGURE 10.20
Window for selection of component on "Design Specification."

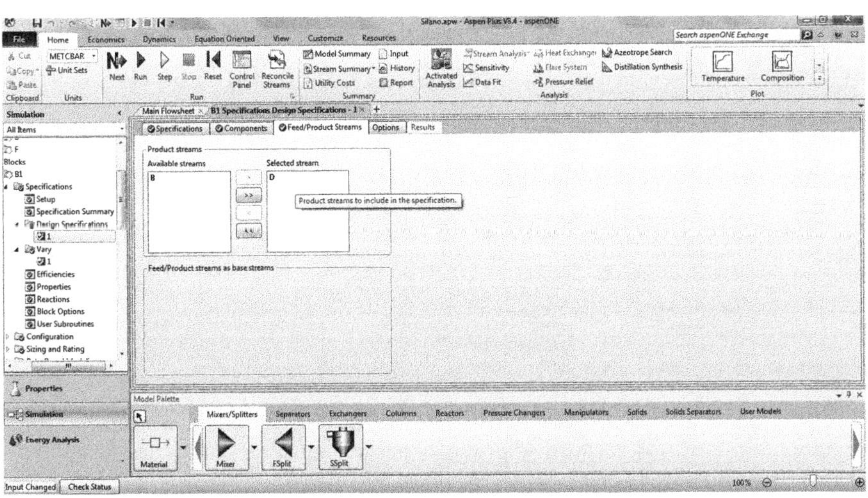

FIGURE 10.21
Window for feed and product current selection on "Design Specification."

adjusted and it is associated with the product on the distillate; the previous steps are described in Figure 10.22.

If we observe Figure 10.21, in the upper left part we have the the Run option to obtain the corresponding results. Once we simulate the process, it is important to consult the results. These results are checked by right clicking the mouse on the column and selecting "Results." Alternatively, we can go to the left part of the window, consulting the folder block of B1 and going to the "Results" folder and then "Profiles." In Figure 10.23, we can see the presented

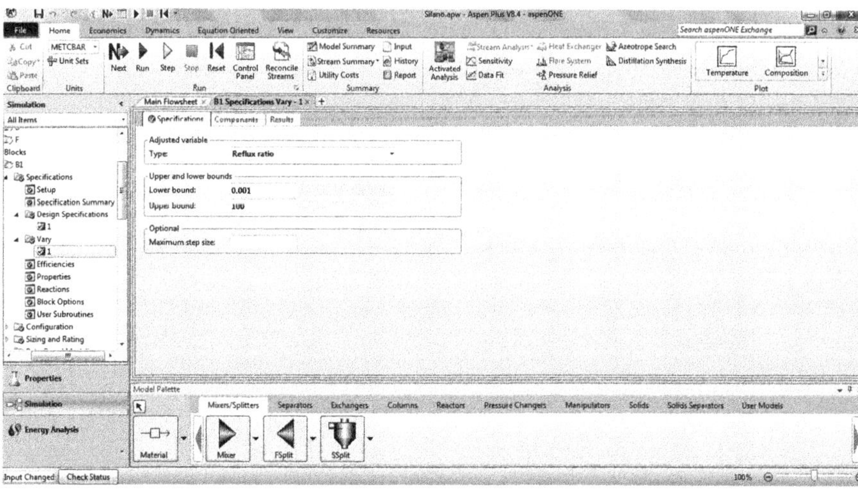

FIGURE 10.22
Window for setting parameters on "Vary."

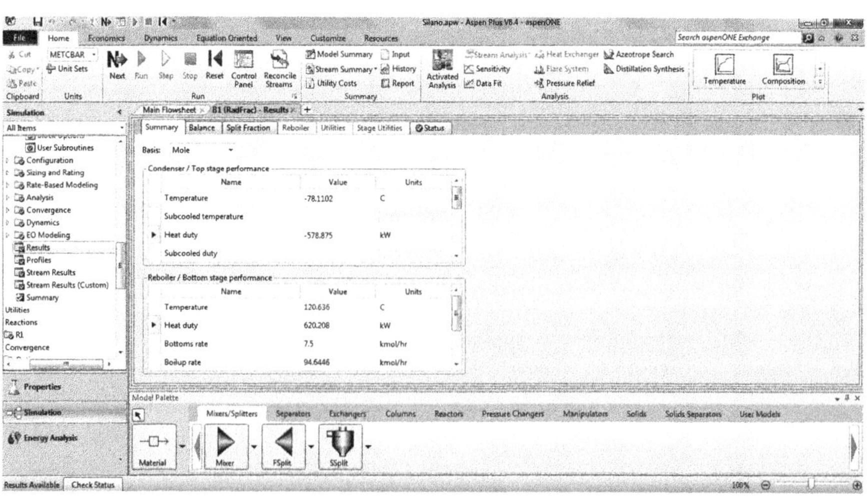

FIGURE 10.23
Window of column results.

results of the column, such as the "Heat Duty" of the condenser and reboiler, as well as other important parameters. In Figures 10.24 and 10.25, we can observe the characteristics of the profile of composition in a quantitative way and the temperature profile on graphic presentation. It is important to indicate that the goal of 0.995 fraction mol of silane was achieved and that the temperature profile in the reactive zone did not exceed 100°C.

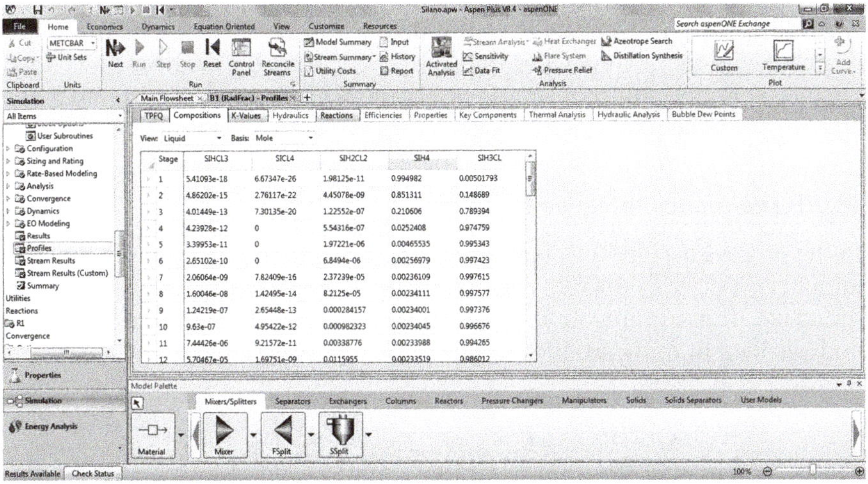

FIGURE 10.24
Window showing the composition profiles.

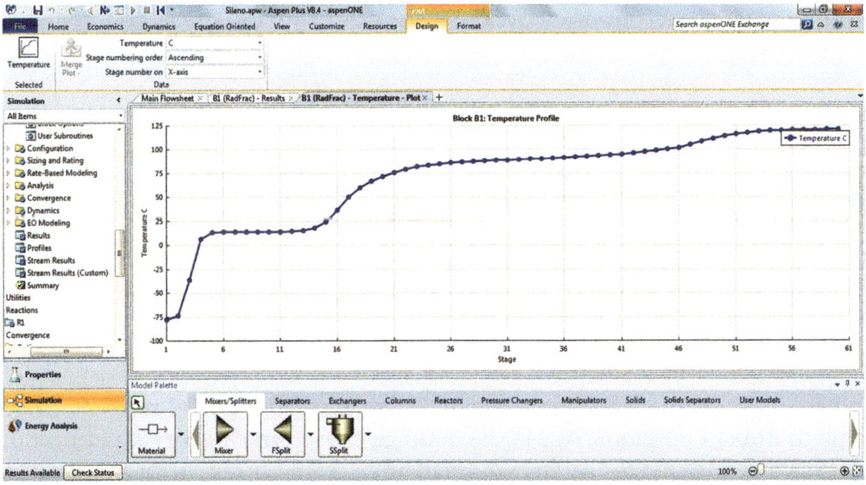

FIGURE 10.25
Temperature graphic window.

10.7 Process Optimization

Previous to the optimization process, the sequence was modeled and simulated rigorously in Aspen Plus using the RadFrac module. Here, it is pretended to obtain a column with good energy consumption reduction, and the configuration that involves the separation of the distillate product, in this case silane, to be 99.5 mol%. The optimized conditions to operate a silane RD column are indispensable to run a silane industry that can effectively compete with the current chlorosilanes derived from the Siemens procedure. Furthermore, the environmental impact should be considered to satisfy the ecological restrictions.

10.7.1 Economic Objective Function

In the process design, the objective function is the minimization of the total annual cost (TAC), which is proportional to the heat duty, services, and column size. The minimization of TAC is subjected to the required recoveries and purities in each product stream, which is stated as follows (Sánchez-Ramírez et al., 2016):

$$\begin{aligned} &\text{Min (TAC)} = \text{f}\,(N_{\text{tn}}, N_{\text{fn}}, N_{\text{rx}}, R_{\text{rn}}, F_{\text{rn}}, D_{\text{cn}}, H_{\text{hd}}) \\ &\text{subject to } \vec{y}_m \geq \vec{x}_m \end{aligned} \tag{10.7}$$

where N_{tn} is the total column stages, N_{fn} is the feed stages in column, N_{rx} is the reaction stages in column, R_{rn} is the reflux ratio, F_{rn} is the distillate fluxes, D_{cn} is the column diameter, H_{hd} is the holdup, and $\vec{y}_m$ and $\vec{x}_m$ are the vectors of the obtained and required purities for the m components, respectively.

10.7.2 Environmental Objective Function

The environmental impact (EI) is measured through the eco-indicator 99, which is based on the methodology of the life-cycle analysis and is stated as follows (Sánchez-Ramírez et al., 2016):

$$\text{Min (Eco-indicator)} = \sum_{b}\sum_{d}\sum_{k \ni K} \partial_d \omega_d \beta_b \alpha_{b,k}, \tag{10.8}$$

where β_b is the total amount of chemical b released per unit of reference flow due to direct emissions, $\alpha_{b,k}$ is the damage caused in category k per unit of chemical b released to the environment, ω_d is a weighting factor for damage in category d, and ∂_d is the normalization factor for damage in category d.

10.7.3 Global Stochastic Optimization

Mainly, the optimization and design of process routes are extremely nonlinear and multivariable problems, with the presence of both continuous and discontinuous design variables. Moreover, the objective functions used as the optimization criteria are potentially nonconvex with the possible presence of local optimums and subject to several constraints.

To optimize the process routes for silane purification, a stochastic optimization method was used, in this case, the differential evolution with tabu list (DETL) method (Sánchez-Ramírez et al., 2016). This method showed that the use of some concepts of the metaheuristic tabu can improve the performance of the DE algorithm. In particular, the tabu list can be used to avoid the revisit of search space by keeping record of the recently visited points, which can avoid unnecessary function evaluations. Based on this fact, the hybrid DETL method is proposed. A complete description of this DETL algorithm is provided by Srinivas and Rangaiah (2007). This optimization approach was implemented using a hybrid platform with Microsoft Excel and Aspen Plus similar to the study of Sánchez-Ramírez et al. (2016).

10.7.4 Results

The Pareto fronts found after the optimization process and three zones in each Pareto front are shown in Figure 10.26. In one part, the most expensive designs are localized but with the minor environmental impact. In contrast, zones where all the designs have the smallest TAC can be observed. Nevertheless, the eco-indicator is the highest. At the middle of both zones, a feasible zone for all processes is located. All those designs accomplish the purities, and the recoveries required and their TACs and eco-indicators 99 are compensated.

Perceptibly, energy savings, leading to new improved designs, are expected. Furthermore, it is clear that, in this analysis, only two objective functions are considered: the economic and the environmental impacts. Nevertheless, an important point of view must be analyzed in future work such as dynamic behavior, i.e., the dynamic properties of this type of process under composition or feed disturbances.

The contour of the Pareto front in Figure 10.26 represents the conflicting targets along the optimization process; in a rough explanation, the blue arrow in the Pareto front consists of designs that preferably include the biggest number of stages (see Tables 10.3 through 10.5), the biggest diameter of column, but minor heat duty. These combinations produced the biggest TAC but the smallest eco-indicator 99. The green arrow consists of designs that preferably include the minor number of stages, the smallest diameter of column, but the biggest heat duty, which produced the lowest TAC but the biggest eco-indicator 99. At the middle of both zones, the red arrow includes

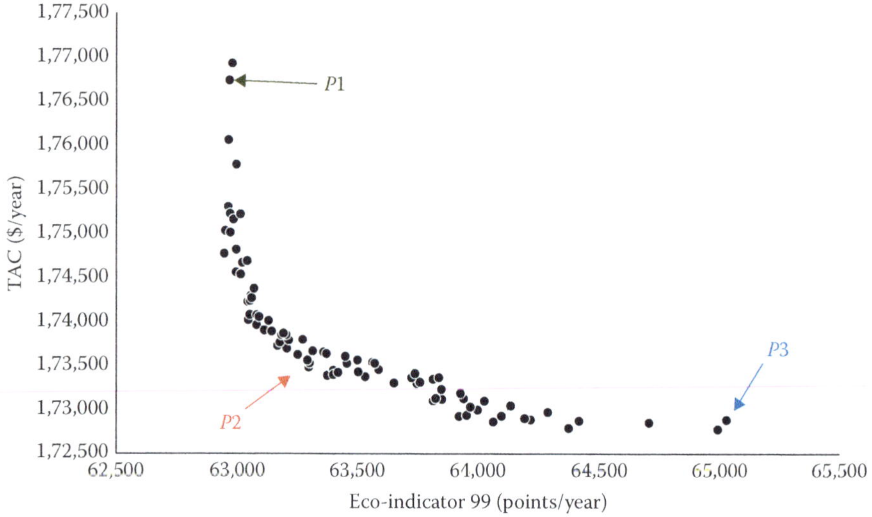

FIGURE 10.26
Pareto front analyzed.

TABLE 10.3
Results of the Low Part of Pareto (P3)

Process Design	
Column Topology	
Number of stages	64
Feed stages	12
Reactive stages	58 (6–64)
Specifications	
Distillate rates (lbmol/h)	5.51187
Reflux ratio	26.6889
Diameter (ft)	0.92499
Holdup (cum)	0.94183
Feed Stream	
Trichlorosilane flow rate (lb/h)	22.0462
Product Streams	
Silane flow rate (lb/h)	5.4844
Silicon tetrachloride flow rate (lb/h)	16.508
Energy Requirements	
Reboiler duty (cal/h)	1061941.6200
Condenser duty (cal/h)	–920715.3650

(Continued)

TABLE 10.3 (*Continued*)

Results of the Low Part of Pareto (P3)

Process Design	
Economic evaluation	
Total annual cost ($/year)	172783.8922
Environmental Impact	
Eco-indicator 99 (points/year)	64995.4554

TABLE 10.4

Results of the Center Part of Pareto (P2)

Process Design	
Column Topology	
Number of stages	80
Feed stages	15
Reactive stages	61 (6–67)
Specifications	
Distillate rates (lbmol/h)	5.4846
Reflux ratio	25.6549
Diameter (ft)	0.9265
Holdup (cum)	0.91400
Feed Stream	
Trichlorosilane flow rate (lb/h)	22.0462
Product Streams	
Silane flow rate (lb/h)	5.4845
Silicon tetrachloride flow rate (lb/h)	16.5084
Energy Requirements	
Reboiler duty (cal/h)	1030091.2300
Condenser duty (cal/h)	–888134.0080
Economic evaluation	
Total annual cost ($/year)	174004.3785
Environmental Impact	
Eco-indicator 99 (points/year)	63051.7303

designs with average variables between both zones, which is reflected in the TAC and eco-indicator values.

Nevertheless, it should be known that the algorithm has explored a different multivariable function since the inclusion of the eco-indicator 99 model, which is loaded with its function, constraints, and so on. In this manner,

TABLE 10.5
Results of the High Part of Pareto (P1)

Process Design	
Column Topology	
Number of stages	80
Feed stages	12
Reactive stages	73 (6–79)
Specifications	
Distillate rates (lbmol/h)	5.5120
Reflux ratio	30.9945
Diameter (ft)	1.0000
Holdup (cum)	0.9483
Feed Stream	
Trichlorosilane flow rate (lb/h)	22.0462
Product Streams	
Silane flow rate (lb/h)	5.4845
Silicon tetrachloride flow rate (lb/h)	16.5084
Energy Requirements	
Reboiler duty (cal/h)	1028597.2100
Condenser duty (cal/h)	–886640.0450
Economic evaluation	
Total annual cost ($/year)	176713.2615
Environmental Impact	
Eco-indicator 99 (points/year)	62967.0524

considering both economic and environmental targets (both conflicting targets), the best solutions can be obtained using the optimization process.

10.8 Conclusions

In this study, we demonstrated the feasibility to produce silane using an RD column, with trichlorosilane as the raw material. This column overcomes the traditional process because production and purification are carried out in a single unit instead of two reactors and four distillation columns. Moreover, all material recycles are avoided, diminishing the energy and equipment requirements. In this single unit, all components obtained were of high purity, exhibiting complete conversion of trichlorosilane to silane. This process was more economic than the conventional process where production and purification were two separated operations. Then, a stochastic global optimization method for the process design of one RD column for the production of silane has been presented. According to the obtained results, the process has shown lowest TAC, but showing a high value for eco-indicator. Moreover, it would be interesting to determine the dynamic behavior of these designs to identify all their process advantages and disadvantages.

References

C.J. Bakay, 1976, Process for Making Silane, US Patent 3,968,199.

A.F.B. Braga, S.P. Moreira, P.R. Zampieri, J.M.G. Bacchin, P.R. Mei, 2008, New processes for the production of solar-grade polycrystalline silicon: A review, *Sol. Energy Mater. Sol. Cells*, 92, 418.

W. Breneman, 1983, Process for Making High Purity Silane, Union Carbide Corporation. Patent number: US4676967.

L. Coleman, 1982, Process for the Production of Ultrahigh Purity Silane with Recycle from Separation Columns, Union Carbide Corporation, US 4,340,574.

A.J. Frings, 2000, Continuous catalytic process for the production of dichlorosilane, in *Proceedings of Silicon for the Chemical Industry V*, Tromsø, Norway.

X. Huang, W.J. Ding, J.M. Yan, W.D. Xiao, 2013, Reactive distillation column for disproportionation of trichlorosilane to silane: Reducing refrigeration load with intermediate condensers, *Ind. Eng. Chem. Res.*, 52, 6211.

K. Inoue, 1988, A Simple Process for the Production of Monosilane, Mitsui Toatsu Chemicals, Inc., JP01317114.

M. Jun-Jian, S.C. Chen, Q. Ke-Qiang, 2007, Thermodynamic study on production of multicrystalline silicon by Siemens process, *Chinese J. Inorg. Chem.*, 23(5), 795–801.

R. Matthes, R. Schork, H.J. Vahlensieck, 1988, Method for the Preparation of Dichlorosilane, Hüls AG, EP 0474265.

A. Morales-Acevedo, G. Casados-Cruz, 2013, Forecasting the development of different solar cell technologies, *Int. J. Photoenergy*, 1–5.

D. Muller, G. Ronge, J.S. Fer, H.J. Leimkuhler, 2002, Development and Economic Evaluation of a Reactive Distillation Process for Silane Production, Distillation and Adsorption: Integrated Processes, Bayer AG, D-51368. *Proceedings of the International Conference on Distillation & Absorption*, Baden-Baden, Germany, pp. 4–11.

S. Pizzini, M. Acciarri, S. Binetti, 2005, From electronic grade to solar grade silicon: Chances and challenges in photovoltaics, *Phys. Status Solidi A*, 202, 2928–2942.

E. Sánchez-Ramírez, J.J. Quiroz-Ramírez, J.G. Segovia-Hernández, S. Hernández, J.M. Ponce-Ortega 2016, Economic and environmental optimization of the biobutanol purification process, *Clean Technol. Environ. Policy*, 18(2), 395–411.

M. Srinivas, G.P. Rangaiah, 2007, Differential evolution with TL for solving nonlinear and mixed-integer nonlinear programming problems, *Ind. Eng. Chem. Res.*, 46, 7126–7135.

O.F. Werner, M. Trygve, H. Arve, M. Morten, K. Hallgeir, 2013, Production of silicon from SiH_4 in a fluidized bed, Operation and results. Report Institute for Energy Technology, Solar Institute 18, Kjeller, Norway, and Telemark University College, Porsgrunn, Norway.

C. Wolden, J. Kurtin, J. Baxter, I. Repins, S. Shaheen, J. Torvik, 2011, Rockett photovoltaic manufacturing: Present status and future prospects, *J. Vac. Sci. Technolo. A.*, 29(3), 030801–030816.

M. Yamada, S. Ishii, T. Nakajima, 1984, Method for Continuous Production of Silanes, Denki Kagaku Kogyo Kabushiki Kaisha, JP121110.

K. Zweibel, J. Mason, V. Fthenakis, 2008, Solar grand plan: Solar as a solution, *Sun & Wind Energy*, 4, 112–117.

Index

A

Acetone–butanol–ethanol (ABE) fermentation, 169
  butanol production by, 171
  limitations of, 170
  prediction of, 178
Active set strategy, 33–36
Age of solution, 50
Ant colony optimization, 49–50
Aspen Batch Modeler, 57
Aspen Custom Modeler, 57
Aspen Distillation Synthesis, 57
Aspen Energy Analyzer, 57
Aspen Plus Dynamics, 57
Aspen Plus simulation engine, 151
Aspen Polymers, 57
Aspen Rate-based Distillation, 57
Augmented Lagrangian function, 32

B

Binary interaction parameters, 65, 80
Biobutanol production process, 169–171
  acetone–butanol–ethanol (ABE) species, 178
  description of, 171–172
  fermentation, 176–177
  kinetic model equations, 174–176
  optimization process, 186–188
  optimization results, 188–190
  UNIQUAC model, 185
Block variable *(Block Var)*, 83, 90, 102

C

Calling subroutine, in Aspen Plus, 158
Chemical process plant, 55
Codification, in genetic algorithm, 42
Components-Specifications sheet, 64, 79, 93
Computer-aided simulation, advantages of, 55–56
Conjugate gradient method, 26
Constrained objective functions, 7
Constraint method, 12
Continuity, 19–24
Continuous function, 19
Conventional production of silane, 195
Convex functions, 21
Convexity, 19–24

D

DE method, *see* Differential evolution method
Deterministic optimization
  continuity and convexity, 19–24
  equality- and inequality-constrained optimization, 32
  single-variable, 15–18
  software for, 36
  unconstrained optimization, 24–25
DETL method, *see* Differential evolution with tabu list method
Dichlorosilane, 195, 198, 200
Differential evolution (DE) method, 43–45
  block diagram for, 46
Differential evolution with tabu list (DETL) method, 213
Di-n-pentyl ether (DNPE), 146–149
  catalytic distillation process for, 153

E

Eco-indicator, 99, 187, 189
Economic objective function, 212
Environmental objective function, 212
Enzymatic hydrolysis, 171
Equality-constrained optimization, 25–26
  active set strategy, 33–36
  generalized reduced gradient method, 29–31
  and inequality, 32–33
  method of Lagrange multipliers, 26–29

"Error accumulators", 58
Excel-Aspen, adding macro in, 136

F

Fantastic harmony, 51
Feasible region, 7
Filling up *User2,* in Aspen Plus, 137
Flash drum, optimization of, 76
 binary interaction parameters, 80
 constraint for, 89
 define sheet for, 87
 initial simulation, 78–82
 objective function, 88
 procedure for, 86
 results for, 91
 sensitivity analysis, 83–85
 Specifications sheet for, 81
 streams results, 82
 thermodynamic method for, 79
 Vary sheet for, 84, 90
 water mole fraction, 85
FORTRAN code, variables in, 144–145
FORTRAN expression, 61
FORTRAN subroutine, 125, 141

G

GAs, *see* Genetic algorithms (GAs)
Generalized reduced gradient method, 29
Genetic algorithms (GAs), 42–43
 block diagram for, 45
Genetic operators, 44
Genotype, 42
Global stochastic optimization, 213
Glucose consumption, metabolic pathway for, 170
Gradient-based methods, 24–25, 26
Gray-box approach, 40

H

Harmony search, 50–51
Heat exchanger, 62
 binary interaction parameters, 65
 define sheet for, 75, 76
 input data for, 66
 measured variable for, 74
 nonrandom two-liquid model, 63
 optimization procedure, 68
 Sensitivity subfolder, 69
 sequential quadratic programming, 73
 Specifications sheet for, 67
 Tabulate sheet, 71
 thermodynamic method, 64
 Vary sheet, 70
 water flow rate, 72
Hessian matrix, 20

I

Initial design, silane production, 200, 201
 column data, 204, 205
 column flow lines, 203
 column results, 210
 "Design Specification", 208, 209
 reaction kinetic parameters, 206
 reactive stages, 207
 temperature graphic, 211
 thermodynamic model, 201, 202

K

Karush–Kuhn–Tucker multipliers, 32
Kinetic model, 91, 95, 173–176
Kinetics subroutine, 143

L

Lagrange multipliers methods, 26–29
Lignocellulosic biomass, 170
Linear programming (LP), 5

M

Mathematical modeling, 1–3
Mathematical programming, 15
Mathematic optimization, type of, 15
MATLAB, 107, 108, 110, 131
 creating function, 110–111
 linking Aspen Plus with, 109
 optimization toolbox, 108, 109, 118
Membrane area, 133, 137

Meta-heuristic optimization, 39
Methanol, molar flow rate of, 82, 83, 86, 90
Methyl tert-butyl ether (MTBE) production, 159
binary interaction parameters, 160
kinetic parameters for, 164
reaction, 163
streams results, 165
Microsoft Excel, 109
creating subroutine using Visual Basic, 111
input data in, 115
sheet "Results", 117, 119
variable explorer, 116
Microsoft excel, creating macro in, 134
MILP, *see* Mixed-integer linear programming
Mixed-integer linear programming (MILP), 5
MLevalstring, 135
MLPutMatrix, 135
Model Analysis Tools folder, 68, 83, 85, 87
Modular process simulators, 58
Mono-objective optimization, 118–119
MTBE, *see* Methyl tert-butyl ether (MTBE)
Multi-objective optimization, 8–11, 119–122, 188
constraint method, 12
weighted sum method, 11–12

N

Newton's method, 26
NLP, *see* Nonlinear programming
Noncontinuous function, 20
Nonconvex functions, 21
Nondominated solution, 10
Nonlinear optimization, 24
Nonlinear programming (NLP), 5, 15
Nonrandom two-liquid (NRTL) model, 63, 78

O

Objective & Constraints sheet, 75, 86, 89, 103
Objective function, 5–7
Optimal control problems, 15
Optimization, 1
classification of, 4–5
constraints, 7–8
mathematical modeling and, 1
multiobjective, 8
objective function, 5
process, 12, 186
results, 188
sensitivity analysis, 68–72
sequential quadratic programming, 73–76
Optimization subfolder, 73, 85, 88, 101
Optimization toolbox, 108, 118

P

ParEnt, 133
Pareto fronts, 10, 213–216
Partial input conditions, of distillation module, 154
Particle swarm optimization, 50
Peng–Robinson equation, 200
Pheromone value, 50
Polycrystalline silicon, 194
Pressure input conditions, of distillation module, 155
Process optimization, 12–14
Process simulation, 55–56
characteristics of, 56–58

R

RadFrac module, 143, 160–162
Reactive distillation (RD), 196
economic potential of, 198, 199
Reactive distillation column
with default kinetic model, 159–165
optimization of, 152
Redlich–Kwong equation, 63

S

Saccharification, 172
Sensitivity analysis tool, in Aspen Plus, 61
Sequential modular simulation, 58–59

Sequential quadratic programming (SQP), 62, 73–76
  optimization, 85–91, 101–106
Silane production process, 193–194
  conventional process, 195–196
  economic potential of reactive distillation, 198
  initial design, *see* Initial design, silane production
  process optimization, 212
  reactive distillation, 196–198
  thermodynamics and kinetic model, 198–200
Silicon tetrachloride, 194, 198, 200
Simulated annealing approach, 47
Simulation
  package, components of, 58
  purpose of, 55
Simultaneous saccharification and fermentation (SSF), 171
Single-variable deterministic optimization, 15–18
SQP, *see* Sequential quadratic programming
SSF, *see* Simultaneous saccharification and fermentation
Steepest descent method, 26
Stochastic optimization, 39
  ant colony optimization, 49–50
  with constraints, 40–41
  differential evolution, 43–45
  genetic algorithms, 42–43
  harmony search, 50–51
  linking Aspen Plus with, 109–110
  particle swarm optimization, 50
  simulated annealing, 47–48
  software for, 107–108
  tabu search, 45–47
  vs. deterministic optimization, 39–40
Stochastic programming, 39
Stoichiometric reactor, 127
  reactor specification of, 128
  stream and reactor results for producing butanol in, 129
Stream input conditions, of distillation module, 155
Subroutine in Microsoft Excel, 111

T

Tabu search, 45–47
Thermodynamic method, 63, 64, 79, 93
Three-phase input conditions, of distillation module, 156
Total annual cost (TAC), 186, 189, 190, 212, 213
Trichlorosilane, proportional decomposition of, 195, 196, 197, 200
Tubular reactor,optimization of, 91
  activation of reaction, 98
  flow rate of methyl lactate, 99
  flowsheet for, 94
  initial simulation, 92
  kinetic parameters of, 97
  Reactions sheet, 96
  sensitivity analysis, 99–101
  sequential quadratic programming, 101–106
  Specifications sheet, 95
  thermodynamic method for, 93

U

Unconstrained optimization, 24–25
UNIFAC model, 92
UNIQUAC model, 185
User2, 130
User-defined block models
  importance of, 125–126
  linking Microsoft Excel and MATLAB, 130–138
  previous work and loading, 126–130
User kinetic model, 141
  in Aspen Plus, 141–143
  di-n-pentyl ether executable code, 146–149
  loading, 149–152
  optimization of reactive distillation column, 152–159
  reactive distillation column, 159–166
  variables in FORTRAN code, 144–145
User kinetics subroutines, in Aspen Plus, 143
User model, 126, 141–143, 150
Userxltemplate file, 131

V

Vapor-liquid equilibrium, 79, 93
Variable explorer, 116
Vector ParReal, 133
Visual basic
  input data in Microsoft Excel, 115
  subroutine inside Microsoft Excel, 111, 114
  variable explorer, 116

W

Weighted sum method, 11

X

Xylose consumption, metabolic pathway for, 170

#0001 - 140817 - C9 - 234/156/13 [15] - CB - 9781498785105